1   Chapter 1: Interactions and Motion

Q01:
Solution:

Reasons 1, 3, and 4 are true. Reason 2 is irrelevant. Reason 5 is correct only if one assumes that the spaceship is indeed effectively infinitely far away from all other sources of gravitational attraction and is thus really only an approximation, but a very good approximation.

Q05:
Solution:

Statements 1 and 5 are correct. Statements 2, 3, and 4 are incorrect.

Q09:
Solution:

Here is a qualitative description of the diagram. During the first 4 minutes, the dots are evenly spaced since the car's speed is constant. During the next 4 minutes, the dots are successively farther apart since the car's speed increases during each minute. During the next 4 minutes, the dots are evenly spaced (approximately twice as far apart as during the first 4 minutes) since the car's speed is now constant once again (but a different constant than before). During the last 4 minutes, the dots are successively closer together since the car's speed is decreasing. The dots must get closer together faster than they got farther apart when the car first accelerated because the speed is decreasing at a greater rate than it increased before.

P15:
Solution:

Extract components by counting gridlines.

(a) $\vec{a} = \langle -4, -3, 0 \rangle$

(b) $\vec{b} = \langle -4, -3, 0 \rangle$

(c) The statement is true. $\vec{a}$ and $\vec{b}$ have the same components, so the two vectors must be equivalent.

(d) $\vec{c} = \langle 4, 3, 0 \rangle$

(e) The statement is true. Each component of $\vec{c}$ is the opposite of the corresponding component of $\vec{a}$ so the actual vectors are opposites.

(f) $\vec{d} = \langle -3, 4, 0 \rangle$

(g) The statement is false because corresponding components of $\vec{c}$ and $\vec{d}$ are not opposites.

P21:
Solution:

The concept of writing a vector as a magnitude multiplying a direction is important and will appear many times in later chapters. It also forces you to think about each part, magnitude and direction, individually.

$$\vec{a} = |\vec{a}| \cdot \hat{a}$$
$$|\vec{a}| = \sqrt{(400 \text{ m/s}^2)^2 + (200 \text{ m/s}^2)^2 + (-100 \text{ m/s}^2)^2}$$
$$|\vec{a}| = 458.3 \text{ m/s}^2$$
$$\hat{a} = \frac{\vec{a}}{|\vec{a}|}$$
$$= \frac{\langle 400, 200, -100 \rangle \text{ m/s}^2}{458.3 \text{ m/s}^2}$$
$$\therefore \vec{a} = 458.3 \text{ m/s}^2 \langle 0.873, 0.436, -0.218 \rangle$$

P23:
Solution:

(a)

$$\vec{r} = \left\langle 3 \times 10^{-10}, -3 \times 10^{-10}, 8 \times 10^{-10} \right\rangle \text{ m}$$

(b)

$$|\vec{r}| = \sqrt{\left(3 \times 10^{-10}\right)^2 + \left(-3 \times 10^{-10}\right)^2 + \left(8 \times 10^{-10}\right)^2} \text{ m}$$
$$= 9.1 \times 10^{-10} \text{ m}$$

(c)

$$\hat{r} = \frac{\vec{r}}{|\vec{r}|}$$
$$= \frac{\left\langle 3 \times 10^{-10}, -3 \times 10^{-10}, 8 \times 10^{-10} \right\rangle \text{ m}}{9.1 \times 10^{-10} \text{ m}}$$
$$= \langle 0.33, -0.33, 0.88 \rangle$$

P27:
Solution:

A helpful hint is to remember that the notation $\vec{r}_{AB}$ is *the position of A relative to B*, which is equivalent to saying *stand at B and tell me how to get to A*. Then you have simply $\vec{r}_{AB} = \vec{r}_A - \vec{r}_B$, with the subtraction done in the order in which the indices appear.

(a)

$$\vec{r}_{\text{planet,star}} = \vec{r}_{\text{planet}} - \vec{r}_{\text{star}}$$
$$= \left\langle -4 \times 10^{10}, -9 \times 10^{10}, 6 \times 10^{10} \right\rangle \text{ m} - \left\langle 6 \times 10^{10}, 8 \times 10^{10}, 6 \times 10^{10} \right\rangle \text{ m}$$
$$= \left\langle -10 \times 10^{10}, -17 \times 10^{10}, 0 \right\rangle \text{ m}$$

(b)

$$\vec{r}_{\text{star,planet}} = -\vec{r}_{\text{planet,star}}$$
$$= \left\langle 10 \times 10^{10}, 17 \times 10^{10}, 0 \right\rangle \text{m}$$

P31:

Solution:

(a)

$$\vec{v}_{\text{avg}} = \frac{\Delta \vec{r}}{\Delta t}$$
$$= \frac{\vec{r}_f - \vec{r}_i}{\Delta t}$$
$$= \frac{\langle -0.202, 0.054, 0.098 \rangle \, \text{m} - \langle 0.2, -0.05, 0.1 \rangle \, \text{m}}{2 \times 10^{-6} \, \text{s}}$$
$$= \frac{\langle -0.402, 0.104, -0.002 \rangle \, \text{m}}{2 \times 10^{-6} \, \text{s}}$$
$$= \left\langle -2.01 \times 10^{5}, 5.2 \times 10^{4}, -1 \times 10^{3} \right\rangle \text{m/s}$$

(b) Average speed is not always equal to the magnitude of average velocity unless the motion is linear. We can proceed with this assumption.

$$|\vec{v}_{\text{avg}}| = \sqrt{\left(-2.01 \times 10^{5} \right)^2 + \left(5.2 \times 10^{4} \right)^2 + \left(-1 \times 10^{3} \right)^2} \, \text{m/s}$$
$$= 2.08 \times 10^{5} \, \text{m/s}$$

P35:

Solution:

$$\vec{r}_i = \left\langle -3 \times 10^{3}, -4 \times 10^{3}, 8 \times 10^{3} \right\rangle \text{m}$$
$$\vec{r}_f = \left\langle -1.4 \times 10^{3}, -6.2 \times 10^{3}, 9.7 \times 10^{3} \right\rangle \text{m}$$
$$t_i = 18.4 \, \text{s}$$
$$t_f = 21.4 \, \text{s}$$
$$\Delta t = 21.4 \, \text{s} - 18.4 \, \text{s}$$
$$= 3.0 \, \text{s}$$

$$\vec{v}_{\text{avg}} = \frac{\Delta \vec{r}}{\Delta t}$$

$$= \frac{\vec{r}_{\text{f}} - \vec{r}_{\text{i}}}{\Delta t}$$

$$= \frac{\langle -1.4 \times 10^3, -6.2 \times 10^3, 9.7 \times 10^3 \rangle \, \text{m} - \langle -3 \times 10^3, -4 \times 10^3, 8 \times 10^3 \rangle \, \text{m}}{3 \, \text{s}}$$

$$= \frac{\langle 1.6 \times 10^3, -2.2 \times 10^3, 1.7 \times 10^3 \rangle \, \text{m}}{3 \, \text{s}}$$

$$= \langle 5.33 \times 10^2, -7.33 \times 10^2, 5.67 \times 10^2 \rangle \, \text{m}$$

P39:
  Solution:

  (a)

$$\vec{r}_{\text{i}} = \langle 0.02, 0.04, -0.06 \rangle \, \text{m}$$
$$\vec{r}_{\text{f}} = \langle 0.02, 1.84, -0.86 \rangle \, \text{m}$$
$$\Delta t = 2 \times 10^{-6} \, \text{s}$$

$$\vec{v}_{\text{avg}} = \frac{\Delta \vec{r}}{\Delta t}$$

$$= \frac{\vec{r}_{\text{f}} - \vec{r}_{\text{i}}}{\Delta t}$$

$$= \frac{\langle 0.02, 1.84, -0.86 \rangle \, \text{m} - \langle 0.02, 0.04, -0.06 \rangle \, \text{m}}{2 \times 10^{-6} \, \text{s}}$$

$$= \frac{\langle 0, 1.8, -0.8 \rangle \, \text{m}}{2 \times 10^{-6} \, \text{s}}$$

$$= \langle 0, 9 \times 10^5, -4 \times 10^5 \rangle \, \text{m/s}$$

  (b) Now, for this time interval of 5×10^{-6} s, the initial position of the electron is its position at the end of the previous 2×10^{-6} s interval.

$$\vec{r}_{\text{i}} = \langle 0.02, 1.84, -0.86 \rangle \, \text{m}$$
$$\Delta t = 5 \times 10^{-6} \, \text{s}$$
$$\vec{v} = \langle 0, 9 \times 10^5, -4 \times 10^5 \rangle \, \text{m/s}$$

$$\vec{r}_f = \vec{r}_i + \vec{v}\Delta t$$

$$= \langle 0.02, 1.84, -0.86 \rangle \, \text{m} + \left(\left\langle 0, 9 \times 10^5, -4 \times 10^5 \right\rangle \text{m/s} \right) \left(5 \times 10^{-6} \, \text{s} \right)$$

$$= \langle 0.02, 1.84, -0.86 \rangle \, \text{m} + \langle 0, 4.5, -2 \rangle \, \text{m}$$

$$= \langle 0.02, 6.34, -2.86 \rangle \, \text{m}$$

Another way to solve it is to consider the total time interval of $2 \times 10^{-6}\,\text{s} + 5 \times 10^{-6}\,\text{s} = 7 \times 10^{-6}\,\text{s}$. In this case, $\vec{r}_i$ is the electron's position at the beginning of the $2 \times 10^{-6}\,\text{s}$ interval.

$$\vec{r}_i = \langle 0.02, 0.04, -0.06 \rangle \, \text{m}$$

$$\Delta t = 5 \times 10^{-6} \, \text{s}$$

$$\vec{r}_f = \langle 0.02, 0.04, -0.06 \rangle \, \text{m} + \left(\left\langle 0, 9 \times 10^5, -4 \times 10^5 \right\rangle \text{m/s} \right) \left(7 \times 10^{-6} \, \text{s} \right)$$

$$= \langle 0.02, 0.04, -0.06 \rangle \, \text{m} + \langle 0, 6.3, -2.8 \rangle \, \text{m}$$

$$= \langle 0.02, 6.34, -2.86 \rangle \, \text{m}$$

which agrees with the same answer obtained using the $5 \times 10^{-6}\,\text{s}$ time interval.

P43:
Solution:

(a)

$$\vec{v}_{\text{avg AB}} = \frac{\Delta \vec{r}}{\Delta t}$$

$$= \frac{\vec{r}_B - \vec{r}_A}{\Delta t}$$

$$= \frac{\langle 22.3, 26.1, 0 \rangle \, \text{m} - \langle 0, 0, 0 \rangle}{1.0\,\text{s} - 0.0\,\text{s}}$$

$$= \langle 22.3, 26.1, 0 \rangle \, \text{m/s}$$

(b) From $t = 1.0\,\text{s}$ to $t = 2.0\,\text{s}$, assuming it travels with a constant velocity of $\langle 22.3, 26.1, 0 \rangle$ m/s,

$$\vec{r}_f = \vec{r}_i + \vec{v}_{\text{avg}}\Delta t$$

$$= \langle 22.3, 26.1, 0 \rangle \, \text{m} + \left(\langle 22.3, 26.1, 0 \rangle \, \text{m/s} \right) \left(2.0\,\text{s} - 1.0\,\text{s} \right)$$

$$= \langle 22.3, 26.1, 0 \rangle \, \text{m} + \langle 22.3, 26.1, 0 \rangle \, \text{m}$$

$$= \langle 44.6, 52.2, 0 \rangle \, \text{m}$$

(c) $\vec{r}$ at point C is $\langle 40.1, 38.1, 0 \rangle$ m which is not the same as what we predicted. We assumed constant velocity when making our prediction; however, in reality the velocity was not constant, but was decreasing in both the x and y directions. An approximation of constant velocity is only valid for small time intervals. For this projectile, $\Delta t = 1.0\,\text{s}$ was not a small enough time interval to reasonably assume constant velocity.

P49:

Solution:

$$\vec{p} = \langle 4, -5, 2 \rangle \, \text{kg} \cdot \text{m/s}$$
$$|\vec{p}| = \sqrt{(4)^2 + (-5)^2 + (2)^2} \, \text{kg} \cdot \text{m/s}$$
$$= 6.7 \, \text{kg} \cdot \text{m/s}$$

P53:

Solution:

It helps to sketch the velocity vector for the basketball before and after it hits the floor.

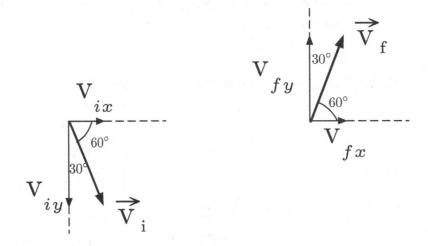

The angle of the vector with the $+y$ axis is $30°$, and the angle with the $+x$ axis is $60°$. The vector's components can be easily calculated using the cosine of each of these angles. Thus

$$\vec{v}_{ix} = |\vec{v}_i| \cos(60°) = 2.5 \, \text{m/s}$$
$$\vec{v}_{iy} = -|\vec{v}_i| \cos(30°) = -4.33 \, \text{m/s}$$

After the ball bounces, $v_{fx} = v_{ix} = 2.5 \, \text{m/s}$ and $v_{fy} = -v_{iy} = +4.33 \, \text{m/s}$. The change in velocity is

$$\Delta \vec{v} = \qquad\qquad\qquad\qquad\qquad\qquad \vec{v}_f - \vec{v}_i$$
$$= \qquad\qquad \langle 2.5, 4.33, 0 \rangle \, \text{m/s} - \langle 2.5, -4.33, 0 \rangle \, \text{m/s}$$
$$= \qquad\qquad\qquad\qquad\qquad\qquad \langle 0, 8.66, 0 \rangle \, \text{m/s}$$

Using the low speed approximation for momentum, the change in momentum is

$$\Delta \vec{p} \approx m \Delta \vec{v}$$
$$\approx (0.57 \, \text{kg})(\langle 0, 8.66, 0 \rangle \, \text{m/s})$$
$$\approx \langle 0, 4.1, 0 \rangle \, \text{kg} \cdot \text{m/s}$$

P59:

    Solution:

$$m = 0.17\,\text{kg}$$
$$\vec{p} = \langle 0, 0, -6.3 \rangle\,\text{kg} \cdot \text{m/s}$$
$$\vec{r}_f = \langle 0, 0, -26 \rangle\,\text{m}$$
$$\Delta t = 0.4\,\text{s}$$
$$\vec{r}_i = ?$$

Since $|\vec{v}| << c$, then $\vec{p} \approx m\vec{v}$.

$$\vec{v} \approx \frac{\vec{p}}{m} = \frac{\langle 0, 0, -6.3 \rangle\,\text{kg} \cdot \text{m/s}}{0.17\,\text{kg}}$$
$$\approx \langle 0, 0, -37.06 \rangle\,\text{m/s}$$

In this case we want to find the initial position (i.e. the position before the 0.4 s time interval). Use the position update equation and solve for the initial position.

$$\vec{r}_f = \vec{r}_i + \vec{v}\Delta t$$
$$\vec{r}_i = \vec{r}_f - \vec{v}\Delta t$$
$$= \langle 0, 0, -26 \rangle\,\text{m} - \left(\langle 0, 0, -37.06 \rangle\,\text{m/s} \right)(0.4\,\text{s})$$
$$= \langle 0, 0, -11 \rangle\,\text{m}$$

Since the velocity was in the z-direction only, then the x-position and y-position did not change.

P63:

    Solution:

$$m_{\text{electron}} = 9.11 \times 10^{-31} \text{kg}$$

$$|\vec{p}| = \gamma m |\vec{v}|$$

$$= \frac{m |\vec{v}|}{\sqrt{1 - \frac{|\vec{v}|^2}{c^2}}}$$

$$= \frac{m (0.9999c)}{\sqrt{1 - \frac{(0.9999c)^2}{c^2}}}$$

$$= \frac{m (0.9999c)}{\sqrt{1 - (0.9999)^2}}$$

$$= \frac{\left(9.11 \times 10^{-31} \text{kg}\right) (0.9999) \left(3 \times 10^8 \text{m/s}\right)}{\sqrt{1 - (0.9999)^2}}$$

$$= 1.93 \times 10^{-20} \text{kg} \cdot \text{m/s}$$

CP69:

Solution:

(a) This program is one possible solution to the problem.

```
from __future__ import division
from visual import *

#length of the side of a cube
a=6

#radius
r=0.5

sphere(pos=(a/2,a/2,a/2), radius=r, color=color.yellow)
sphere(pos=(a/2,a/2,-a/2), radius=r, color=color.yellow)
sphere(pos=(a/2,-a/2,-a/2), radius=r, color=color.yellow)
sphere(pos=(-a/2,-a/2,-a/2), radius=r, color=color.yellow)
sphere(pos=(a/2,-a/2,a/2), radius=r, color=color.magenta)
sphere(pos=(-a/2,a/2,a/2), radius=r, color=color.magenta)
sphere(pos=(-a/2,-a/2,a/2), radius=r, color=color.magenta)
sphere(pos=(-a/2,a/2,-a/2), radius=r, color=color.magenta)
```

It produces the following screen capture.

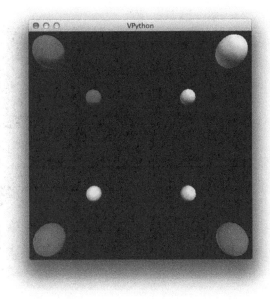

(b) It is the same program as in part (a), but with one additional line of code to create the arrow.

arrow(pos=vector($-a/2,-a/2,-a/2$), axis=vector($a/2,a/2,a/2$)$-$vector($-a/2,-a/2,-a/2$), color=color.cyan)

Rather than calculating the axis of the arrow by hand or trial and error, it is easiest to use VPython to subtract the position of one sphere from the position of another sphere on the diagonally opposite side of the cube. It helps to think "final minus initial" when you subtract positions of the spheres to get the correct axis for the arrow. (In P72, you will learn now to name the objects so that you can easily refer to their attributes rather than retyping their values.)

```
from __future__ import division
from visual import *

#length of the side of a cube
a=6

#radius
r=0.5

sphere(pos=(a/2,a/2,a/2), radius=r, color=color.yellow)
sphere(pos=(a/2,a/2,-a/2), radius=r, color=color.yellow)
sphere(pos=(a/2,-a/2,-a/2), radius=r, color=color.yellow)
sphere(pos=(-a/2,-a/2,-a/2), radius=r, color=color.yellow)
sphere(pos=(a/2,-a/2,a/2), radius=r, color=color.magenta)
sphere(pos=(-a/2,a/2,a/2), radius=r, color=color.magenta)
sphere(pos=(-a/2,-a/2,a/2), radius=r, color=color.magenta)
sphere(pos=(-a/2,a/2,-a/2), radius=r, color=color.magenta)


arrow(pos=vector(-a/2,-a/2,-a/2), axis=vector(a/2,a/2,a/2)-vector(-a/2,-a/2,-a/2),
    color=color.cyan)
```

It produces the following screen capture.

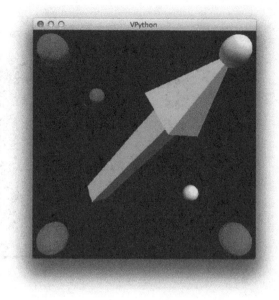

CP75:

Solution:

(a) Examine the program below. It's useful to define variables for the length of the side of a box, the length of an axis, and the number of boxes along the axis. The space between boxes can be calculated from the length of the axis and the number of boxes along the axis.

```
from __future__ import division
from visual import *

#length of the axis
a=1

#width of a box
w=a/100

#Number of boxes on an axis
n=20

#x location of a box
x=-a/2

#space between boxes
dx=a/(n-1)

#boxes on x-axis
while x<a/2:
    box(pos=(x,0,0),size=(w,w,w),color=color.yellow)
    x=x+dx

#y location of a box
y=-a/2
```

```
#space between boxes
dy=dx

#boxes on y-axis
while y<a/2:
    box(pos=(0,y,0),size=(w,w,w),color=color.magenta)
    y=y+dy

#z location of a box
z=-a/2

#space between boxes
dz=dx

#boxes on z-axis
while z<a/2:
    box(pos=(0,0,z),size=(w,w,w),color=color.cyan)
    z=z+dz
```

The output is shown below.

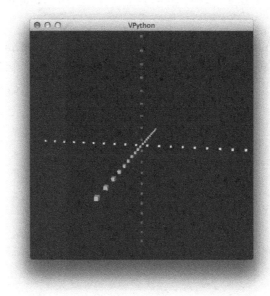

(b) There are many ways to solve this problem. In the sample program below, we draw n boxes on an axis. So for three axes, our `while` loop requires $3n$ iterations. Using an `if` statement, we check to see how many boxes have been drawn, and we use this value to determine the axis on which we will draw the next box. It's important to reset the value of x for the position of the first box whenever starting a new axis.

```
from __future__ import division
from visual import *

#length of the axis
a=1

#width of a box
w=a/100
```

```
#Number of boxes on an axis
n=20

#x location of a box
x=-a/2

#space between boxes
dx=a/(n-1)

#There is a total of 3*n boxes.
#After n boxes on the x-axis, then create boxes on the y-axis.
#After another n boxes on the y-axis, then create boxes on the z-axis
boxnum=1
while boxnum<3*n+1:
    #x axis
    if(boxnum<n+1):
        box(pos=(x,0,0),size=(w,w,w),color=color.yellow)
    #y axis
    elif(boxnum<2*n+1 and boxnum>n):
        box(pos=(0,x,0),size=(w,w,w),color=color.magenta)
    #z axis
    else:
        box(pos=(0,0,x),size=(w,w,w),color=color.cyan)
    #have to reset the value of x to -a/2 when starting a new axis
    if(boxnum==n or boxnum==2*n):
        x=-a/2

    x=x+dx
    boxnum=boxnum+1
```

2 Chapter 2: The Momentum Principle

Q07:
Solution:

$\Delta \vec{p}$ is in the direction of the net force on the puck. Thus, $\Delta \vec{p}$ is in the direction of arrow (e).

P11:
Solution:

$$t_i = 16\,\text{s}$$
$$t_f = 16.2\,\text{s}$$
$$m = 4\,\text{kg}$$
$$\vec{v}_i = \langle 9, 29, -10 \rangle\,\text{m/s}$$
$$\vec{v}_f = \langle 18, 20, 25 \rangle\,\text{m/s}$$
$$\vec{F}_{net} = ?$$

$$
\begin{aligned}
\vec{F}_{net} &= \frac{\Delta \vec{p}}{\Delta t} \\
&= \frac{m \Delta \vec{v}}{\Delta t} \\
&= \frac{m\left(\vec{v}_f - \vec{v}_i\right)}{\Delta t} \\
&= \frac{(4\,\text{kg})\left(\langle 18, 20, 25 \rangle\,\text{m/s} - \langle 9, 29, -10 \rangle\,\text{m/s}\right)}{(16.2\,\text{s} - 16.0\,\text{s})} \\
&= \frac{(4\,\text{kg})\left(\langle 9, -9, 35 \rangle\,\text{m/s}\right)}{0.2\,\text{s}} \\
&= \langle 180, -180, 700 \rangle\,\text{N}
\end{aligned}
$$

P15:
Solution:

$$m = 5\,\text{kg}$$
$$\vec{F}_{net} = \langle 29, -15, 40 \rangle\,\text{N}$$
$$\Delta t = 4\,\text{s}$$
$$\vec{v}_f = \langle 114, 74, 112 \rangle\,\text{m/s}$$
$$\vec{v}_i = 0$$

Define the system to be the rock.

$$\vec{F}_{net} = \frac{\Delta \vec{p}}{\Delta t}$$

$$= \frac{m \Delta \vec{v}}{\Delta t}$$

$$\vec{F}_{net} \Delta t = m \Delta \vec{v}$$

$$\vec{F}_{net} \Delta t = m \vec{v}_f - m \vec{v}_i$$

$$m \vec{v}_i = m \vec{v}_f - \vec{F}_{net} \Delta t$$

$$\vec{v}_i = \vec{v}_f - \frac{\vec{F}_{net}}{m} \Delta t$$

$$= \langle 114, 94, 112 \rangle \, \text{m/s} - \left(\frac{\langle 29, -15, 40 \rangle \, \text{N}}{5 \, \text{kg}} \right) (4 \, \text{s})$$

$$= \langle 114, 94, 112 \rangle \, \text{m/s} - \langle 23.2, -12, 32 \rangle \, \text{m/s}$$

$$= \langle 91, 106, 80 \rangle \, \text{m/s}$$

P19:

Solution:

Basically, we need to deflect the object so that it doesn't collide with Earth. If the object is a comet nucleus, there is a high probability that its orbit is in a different plane than Earth's orbit is in. If the object is not a comet, but an asteroid fragment, then there is a greater probability that its orbit is coplanar, or very nearly coplanar, with Earth's orbit so let's assume this is the situation. Now we have two choices. We can either attempt to deflect the object out of its orbital plane or we can attempt to deflect it away from Earth but in its own orbital plane. Let's assume the latter. We need to give the object a component of momentum (and thus also velocity) perpendicular to its initial trajectory, and we do this by slamming a spacecraft into the object. The object's initial momentum is directed toward Earth, but its final momentum will have an additional component perpendicular to the initial component. Let's neglect Earth's motion during this mission, at least for the moment. Choose the system to be the incoming object and the spacecraft. Let's also assume non-relativistic speeds. Assume further that the spacecraft slams into the object from a direction perpendicular to the object's initial momentum and becomes completely embedded. Finally, let's neglect interactions outside the system.

$$\vec{p}_{sys,final} = \vec{p}_{sys,initial} + \vec{F}_{net,sys} \Delta t$$

$$\vec{p}_{final,craft+object} \approx \vec{p}_{initial,craft} + \vec{p}_{initial,object}$$

$$\left(m_{craft} + m_{object} \right) \vec{v}_{final} \approx m_{craft} \vec{v}_{initial,craft} + m_{object} \vec{v}_{initial,object}$$

$$\vec{v}_{final} \approx \frac{m_{craft} \vec{v}_{initial,craft} + m_{object} \vec{v}_{initial,object}}{\left(m_{craft} + m_{object} \right)}$$

Now we need some reasonable physical values. A Saturn V rocket has a mass of about $3 \times 10^6 \, \text{kg}$, but what about the object? Asteroids vary in composition and density, but Ceres, the largest asteroid, has a density approximately twice that of water, or about $2 \times 10^3 \, \text{kg/m}^3$. The object is approximately spherical, so it's mass must be about $\left(2 \times 10^3 \, \text{kg/m}^3 \right) \times$

$\frac{4}{3}\pi \left(50\,\text{m}^3\right) \approx 1 \times 10^9\,\text{kg}.$

$$\vec{v}_{\text{final}} \approx \frac{\left(3 \times 10^6\,\text{kg}\right) \left\langle 1 \times 10^3, 0, 0 \right\rangle\,\text{m/s} + \left(1 \times 10^9\,\text{kg}\right) \left\langle 0, 3 \times 10^4, 0 \right\rangle\,\text{m/s}}{\left(3 \times 10^6\,\text{kg} + 1 \times 10^9\,\text{kg}\right)}$$

$$\approx \frac{\left\langle 3 \times 10^9, 0, 0 \right\rangle\,\text{kg} \cdot \text{m/s} + \left\langle 0, 3 \times 10^{13}, 0 \right\rangle\,\text{kg} \cdot \text{m/s}}{1.003 \times 10^9\,\text{kg}}$$

$$\approx \frac{\left\langle 3 \times 10^9, 3 \times 10^{13}, 0 \right\rangle\,\text{kg} \cdot \text{m/s}}{1.003 \times 10^9\,\text{kg}}$$

$$\approx \left\langle 3, 3 \times 10^4, 0 \right\rangle\,\text{m/s}$$

This doesn't appear to change the object's velocity (actually the velocity of the object and embedded spacecraft) very much. However, this rendezvous takes place 2.5×10^{11} m from Sun. The object's initial speed (and final speed) are the same as Earth's orbital speed, so they travel about the same distance each day. Will this small deflection be enough to prevent a collision with Earth?

P23:
Solution:

The ball is rolling in the xz-plane. Before you kick it, its velocity and momentum are in the $+z$ direction. Your kick imparts a momentum component in the $-x$ direction, so that immediately after your kick the ball's velocity and momentum have two nonzero components. The net force on the ball is opposite the direction of the momentum. Assume time starts passing immediately after your kick and at that instant, the ball is at the origin. Carry out a step by step calculation as before.

At $t = 0$ s we have $\vec{r} = \langle 0, 0, 0 \rangle$ m, $\vec{v} = \langle -3.023, 0, 2.2 \rangle$ m/s, $m = 0.43$ kg, $\vec{p} = \langle -1.3, 0, 0.946 \rangle$ kg $\cdot$ m/s, $\hat{p} = \langle -0.809, 0, 0.588 \rangle$, and $\vec{F}_{\text{net}} = \langle 0.202, 0, -0.147 \rangle$ N.

With these initial conditions, the following VPython program does the calculations and thus at $t = 1.5$ s we have $\vec{r} \approx \langle -3.83, 0, 2.79 \rangle$ m.

```
from __future__ import division, print_function
from visual import *

m = 0.43
r = vector(0,0,0)
print ("r=",r)
v = vector(-3.023,0,2.2)
print ("v=",v)
p = m*v
print ("p=",p)
phat = norm(p)
print ("phat=",phat)
Fnet = -0.25 * phat
print ("Fnet=",Fnet)
dt = 0.5
t = 0
print ("t=",t,"r=",r,"v=",v)
while (t<01.5):
    p = p + Fnet * dt
    v = p/m
```

```
r = r + (p/m) * dt
t += dt
print ("t=",t,"r=",r,"v=",v)
```

P29:

Solution:

(a) v_x is negative and "increases" toward zero, meaning that it becomes less negative as the cart slows down. There is no graph that demonstrates this motion. Graph (2) is close, but in graph (2), the object reaches zero velocity and then speeds up as it travels to the right.

(b) v_x is positive and increases. This is graph (6).

(c) v_x is negative and constant. This is graph (3).

(d) v_x is negative, decreases in magnitude until $v_x = 0$, then increases and is positive. This is graph (2).

(e) v_x is zero and constant. No graphs depict this motion.

(f) v_x is positive and decreases. No graphs depict this motion. Graph (5) is close, but it shows the object's velocity decreasing to 0, and then the object speeds up in the -x direction.

(g) v_x is positive, decreases to 0, and then becomes more negative (i.e. increases in the -x direction). This is graph (5).

(h) v_x is negative and becomes more negative as the cart speeds up. This is graph (1).

(i) v_x is constant and positive. This is graph (4).

P33:

Solution:

First, sketch a picture of the situation.

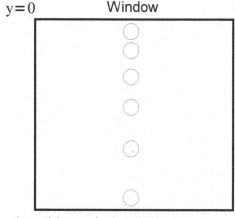

$$y_f = -(0.85)(2.2 \, \text{m}) = -1.87 \, \text{m}$$

P39:

Solution:

First, sketch a picture of the test track. Let's call region A the first half of the track ($0 \le x < L/2$). Let's call region B the last half of the track ($L/2 < x \le L$). Let's call the length of the track L.

The cars have different accelerations, so let's call their x-accelerations a_{1x} and a_{2x} respectively. For constant net force, use these equations

$$v_{x,f} = v_{x,i} + a_x \Delta t$$
$$\Delta x = v_{x,i} \Delta t + \frac{1}{2} a_x (\Delta t)^2$$

Begin with Car 1. You have to analyze its motion during each region (A and B) separately since its acceleration is different in the two regions. In region B, $a_1 = 0$.

$$\text{Car 1, Region A:} \quad \Delta x = v_{x,i} \Delta t + \frac{1}{2} a_x (\Delta t)^2$$
$$= \frac{1}{2} a_x (\Delta t)^2$$
$$\frac{L}{2} = \frac{1}{2} a_1 \Delta t_A$$
$$\Delta t_A = \sqrt{\frac{L}{a_1}}$$
$$v_{x,f} = v_{x,i} + a_x \Delta t$$
$$= a_1 \sqrt{\frac{L}{a_1}}$$
$$v_{x,f} = \sqrt{a_1 L}$$

$$\text{Car 1, Region B:} \quad \Delta x = v_x \Delta t$$
$$\frac{L}{2} = \sqrt{a_1 L} \Delta t_B$$
$$\Delta t_B = \frac{1}{2} \sqrt{\frac{L}{a_1}}$$

Thus, the total time for Car 1 to travel down the track is $\Delta t = \sqrt{\frac{L}{a_1}} + \frac{1}{2}\sqrt{\frac{L}{a_1}} = \frac{3}{2}\sqrt{\frac{L}{a_1}}$.

Now analyze Car 2. It has constant acceleration for the entire length of the track. There's no need to analyze regions A and B separately.

$$\text{Car 2:} \qquad \Delta x = v_{x,i}\,\Delta t + \frac{1}{2}a_x\,(\Delta t)^2$$

$$= \frac{1}{2}a_x\,(\Delta t)^2$$

$$L = \frac{1}{2}a_2\,\Delta t^2$$

$$\Delta t = \sqrt{\frac{2L}{a_2}}$$

$$v_{x,f} = v_{x,i} + a_x\,\Delta t$$

$$= a_2\sqrt{\frac{2L}{a_2}}$$

$$v_{x,f} = \sqrt{2a_2 L}$$

(a) Using these results we can answer the questions. First find the average speed of each car. Use the conventional definition of average speed:

$$\text{average speed} = \frac{\text{distance traveled}}{\text{time elapsed}}$$

Note that this does not generally give the same result as $\left|\vec{v}_{\text{avg}}\right| = \left|\frac{\vec{v}_i + \vec{v}_f}{2}\right|$, except in the case of constant acceleration (such as Car 2).

Both cars travel the distance L in the time Δt. Thus, they have the same average speed, and the ratio of their average speeds is 1.

(b) We have already solved for the total time traveled by each car in terms of the distance L and their accelerations. They reach the end of the track in the same time. Set their times equal to each other and solve for the ratio of accelerations.

$$\Delta t_1 = \Delta t_2$$

$$\frac{3}{2}\sqrt{\frac{L}{a_1}} = \sqrt{\frac{2L}{a_2}}$$

$$\frac{9}{4}\frac{L}{a_1} = \frac{2L}{a_2}$$

$$\frac{a_1}{a_2} = \frac{9}{8}$$

(c) The ratio of their final x-velocities is

$$\frac{v_{x,f1}}{v_{x,f2}} = \frac{\sqrt{a_1 L}}{\sqrt{2a_2 L}}$$

$$= \sqrt{\frac{a_1 L}{2a_2 L}} = \sqrt{\frac{a_1}{2a_2}}$$

$$= \sqrt{\frac{9}{16}} = \frac{3}{4}$$

It is nice to check our work with a VPython program. Using $L = 100\,\text{m}$ and $a_1 = 25\,\text{m/s}^2$, a VPython program produced the following graphs of $x(t)$ and $v_x(t)$ for the two cars. By knowing that $a_1 = \frac{9}{8}a_2$, you should be able to figure out which curves (on each graph) represent Car 1 and Car 2, respectively. From the graphs you should also be able to measure the total time to reach the end of the track and the final x-velocity of each cart. Then you can compare the values to what you calculate from the equations, using $L = 100\,\text{m}$ and $a_1 = 25\,\text{m/s}^2$.

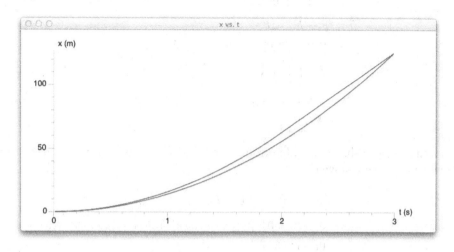

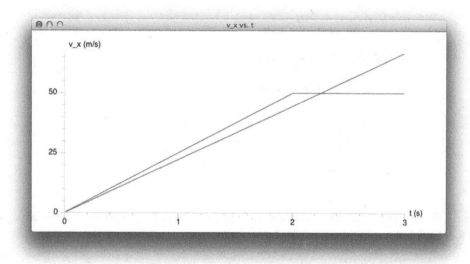

P43:

Solution:

Begin with $\vec{r} = \langle 0.0.0798, 0 \rangle\,\text{m}, \vec{v} = \langle 0.0877, 0 \rangle\,\text{m/s}, \vec{p} = \langle 0, 0.0307, 0 \rangle\,\text{kg} \cdot \text{m/s}, \Delta t = 0.1\,\text{s}$. Actual calculations were done in Python.

Here's the first time step.

$$\vec{F}_{net} = \left\langle 0, -mg - k_s \left(L - L_o\right), 0 \right\rangle$$
$$= \left\langle 0, -\left(0.350\,\text{kg}\right)\left(9.80\,\text{N/kg}\right) - \left(55\,\text{N/m}\right)\left(-0.1502\,\text{m}\right), 0 \right\rangle$$
$$= \left\langle 0, 4.831, 0 \right\rangle \text{N}$$
$$\vec{p} = \vec{p} + \vec{F}_{net}\Delta t$$
$$= \left\langle 0, 0.5138, 0 \right\rangle \text{kg} \cdot \text{m/s}$$
$$\vec{r} = \vec{r} + \left(\frac{\vec{p}}{m}\right)\Delta t$$
$$= \left\langle 0, 0.2266, 0 \right\rangle \text{m}$$

Here's the second time step, using final value from first time step as new initial values. Note that the force changed because the amount of stretch changed.

$$\vec{F}_{net} = \left\langle 0, -mg - k_s \left(L - L_o\right), 0 \right\rangle$$
$$= \left\langle 0, -\left(0.350\,\text{kg}\right)\left(9.80\,\text{N/kg}\right) - \left(55\,\text{N/m}\right)\left(-0.0034\,\text{m}\right), 0 \right\rangle$$
$$= \left\langle 0, -3.243, 0 \right\rangle \text{N}$$
$$\vec{p} = \vec{p} + \vec{F}_{net}\Delta t$$
$$= \left\langle 0, 0.1895, 0 \right\rangle \text{kg} \cdot \text{m/s}$$
$$\vec{r} = \vec{r} + \left(\frac{\vec{p}}{m}\right)\Delta t$$
$$= \left\langle 0, 0.2807, 0 \right\rangle \text{m}$$

CP49:
Solution:

(a) In Problem P48, two programs were written. In P48 (a), the net force on the cart is zero, so the cart's momentum is constant. Graphs of $x\left(t\right)$ and $p_x\left(t\right)$ are shown below.

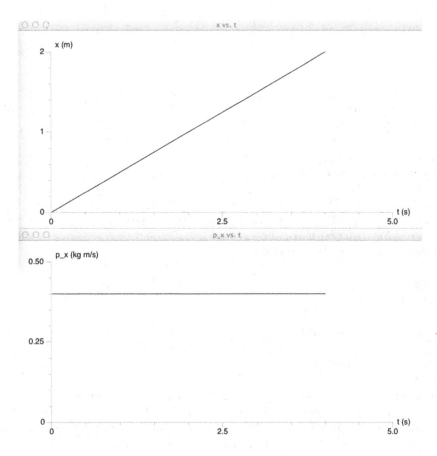

The program used to simulate the cart and plot the graphs is given below.

```
from __future__ import division
from __future__ import print_function
from visual import *
from visual.graph import *

# make a track
# Note that the track is a bit longer than 2 m to accommodate the cart
track = box(pos=vector(1,-0.05,0), size=(2.1,0.05,0.10), color=color.white)

# make a cart
cart = box(pos=vector(0,0,0), size=(0.1,0.04,0.06), color=color.red)

# define cart's mass in SI units
# read the variable name from right to left: mass of the object named "cart"
cart.m = 0.8

# define cart's initial momentum (and initial velocity) in SI units
# read the variable name from right to left: momentum of the object named "cart"
cart.p = cart.m * vector(0.5,0,0)

# define net force on cart
Fnet = vector(0,0,0)

# timestep
```

```
deltat = 0.01

# initial time
t = 0

# initialize x graph
graph1 = gdisplay(x=450, y=0, title='x vs. t', xtitle='t (s)',
                  ytitle='x (m)',xmin=0., xmax=5.0, ymin=0., ymax=2.0)
# initialize the function to be plotted
xvt = gcurve(color=color.green)
# initial p_x graph
graph2 = gdisplay(x=450, y=450, title='p_x vs. t', xtitle='t (s)',
                  ytitle='p_x (m)',xmin=0., xmax=5.0, ymin=0., ymax=0.5)
# initialize the function to be plotted
pxvt = gcurve(color=color.green)

# pause and wait for user to click mouse to start animation
scene.mouse.getclick()

# loop to animate the cart
while 1:
    # limit animation rate
    rate(100)
    # apply momentum principle to update cart's momentum
    cart.p = cart.p + Fnet * deltat
    # update cart's position, assuming Newton's expression for momentum
    cart.pos = cart.pos + (cart.p/cart.m) * deltat
    if cart.x > 2:
        print("cart is at x=",cart.x," at t= ",t)
        break
    # update elapsed time
    t = t + deltat
    # update x graph
    xvt.plot(pos=(t,cart.x))
    # update p_x graph
    pxvt.plot(pos=(t,cart.p.x))
```

The slope of the $x(t)$ graph is positive and constant. The cart starts at the origin and travels to the right with a constant x-velocity (the slope). The graph of $p_x(t)$, shows that the value of p_x is constant.

(b) In P48 (c), the net force on the cart is to the right and the initial velocity is to the left. As a result, the cart travels to the left and slows down until it momentarily stops. Then it travels to the right and speeds up. Graphs of $x(t)$ and $p_x(t)$ are shown below.

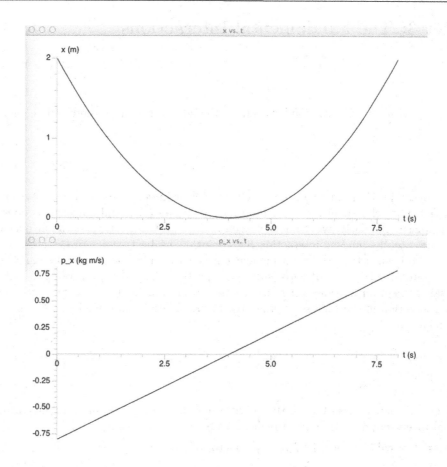

The graph of $x(t)$ shows that the cart starts at $x = 2\,\text{m}$ and travels to the left (negative slope). The slope decreases to zero, so the cart is slowing down. At $x = 0$ ($t = 4\,\text{s}$), the slope is zero, so this is when (and where) the cart momentarily stops. After $t = 4\,\text{s}$, the slope of the curve is positive so the cart travels to the right. Since the slope increases, the cart speeds up as it travels to the right.

The graph of $p_x(t)$ shows that the initial value of p_x is negative. Thus at $t = 0$, the cart is traveling to the left. The absolute value $|p_x|$ decreases to 0 at $t = 4\,\text{s}$. Thus, for $t < 4\,\text{s}$, the cart is slowing down. For $t > 4\,\text{s}$, the value of p_x is positive and increases so the cart speeds up as it travels to the right. The slope of $p_x(t)$ is constant; therefore, the x-component of the net force on the cart is constant.

3   Chapter 3: The Fundamental Interactions

Q03:

Solution:

There is only **one** gravitational interaction between Earth and tennis ball, so there can be only **one** associated magnitude.

Q07:

Solution:

This is a nuclear decay problem, and so is governed by nuclear interactions. Choice of system is extremely important in this problem. We cannot choose system = radium nucleus because the radium nucleus does not exist after the nuclear interaction takes place. Similarly, we cannot choose system = alpha particle + radon nucleus because neither of these particles exists prior to the nuclear interaction happening. The best choice of system is system = all particles. The system's initial state is such that the total momentum is zero (only one stationary particle). Note that in the initial state, the system's total energy is **not** zero (stationary particles have energy). With no net force acting on our chosen system, the total momentum must be zero in both the initial and final states. So if the newly created alpha particle moves in the $+z$ direction in the final state, then the newly created radon nucleus must move in the $-z$ direction so as to conserve momentum.

P11:

Solution:

According to Newton's law of gravitation, the gravitational force on M by m is directly proportional to the product of their masses and inversely proportional to the distance between them squared.

Tripling the mass of m will increase the force by a factor of 3.

Increasing the distance by a factor of 4 will decrease the force by 4^2, making the force smaller by $\left(\frac{1}{4}\right)^2 = \frac{1}{16}$.

Thus the objects will attract with a force

$$F' = \left(\frac{3}{16}\right) F$$

P19:

Solution:

(a) The relative position vector of the star, relative to the planet, is

$$
\begin{aligned}
\vec{r}_{\text{from planet to the star}} &= \vec{r}_{\text{star}} - \vec{r}_{\text{planet}} \\
&= \left\langle -2 \times 10^{11}, 3 \times 10^{11}, 0 \right\rangle \text{m} - \left\langle 5 \times 10^{11}, -2 \times 10^{11}, 0 \right\rangle \text{m} \\
&= \left\langle -7 \times 10^{11}, 5 \times 10^{11}, 0 \right\rangle \text{m}
\end{aligned}
$$

(b) The distance is

$$\left| \vec{r}_{\text{from planet to the star}} \right| = \sqrt{\left(7 \times 10^{11}\, \text{m}\right)^2 + \left(5 \times 10^{11}\, \text{m}\right)^2 + (0\, \text{m})^2}$$

$$= 8.6 \times 10^{11}\, \text{m}$$

(c) The direction of the relative position vector of the star, relative to the planet, is

$$\hat{r}_{\text{from planet to the star}} = \frac{\left\langle -7 \times 10^{11}, 5 \times 10^{11}, 0 \right\rangle\, \text{m}}{8.6 \times 10^{11}\, \text{m}}$$

$$= \left\langle -0.81, 0.58, 0 \right\rangle$$

(d) Newton's law of gravitation is

$$\left| \vec{F}_{\text{grav on planet by star}} \right| = G \frac{m_{\text{planet}}\, m_{\text{star}}}{|\vec{r}|^2}$$

$$= \left(6.6738 \times 10^{-11}\, \text{N} \cdot \text{m}^2/\text{kg}^2\right) \frac{\left(4 \times 10^{24}\, \text{kg}\right)\left(5 \times 10^{30}\, \text{kg}\right)}{\left(8.6 \times 10^{11}\, \text{m}\right)^2}$$

$$= 1.8 \times 10^{21}\, \text{N}$$

(e) Because of the Principle of Reciprocity, the planet and star exert forces of equal magnitude on one another. Thus,

$$\left| \vec{F}_{\text{grav on planet by star}} \right| = \left| \vec{F}_{\text{grav on star by planet}} \right|$$

$$= 1.8 \times 10^{21}\, \text{N}$$

(f) The (vector) force exerted on the planet by the star is

$$\vec{F}_{\text{on the planet by the star}} = \left| \vec{F}_{\text{on the planet by the star}} \right| \hat{r}_{\text{from planet to star}}$$

$$= \left(1.8 \times 10^{21}\, \text{N}\right) \left\langle -0.81, 0.58, 0 \right\rangle$$

$$= \left\langle -1.5 \times 10^{21}, 1.0 \times 10^{21}, 0 \right\rangle\, \text{N}$$

(g) The (vector) force exerted on the star by the planet is

$$\vec{F}_{\text{on the star by the planet}} = \left| \vec{F}_{\text{on the star by the planet}} \right| \left(-\hat{r}_{\text{from planet to star}} \right)$$

$$= \left(1.8 \times 10^{21}\, \text{N}\right) \left\langle 0.81, -0.58, 0 \right\rangle$$

$$= \left\langle 1.5 \times 10^{21}, -1.0 \times 10^{21}, 0 \right\rangle\, \text{N}$$

P23:

Solution:

At an altitude y, the distance from Earth's surface is $R + y$. At this height, the gravitational field is 99% of the field at Earth's surface.

$$\left|\vec{g}_{\text{at surface}}\right| = G\frac{m_{\text{Earth}}}{R^2}$$

$$\left|\vec{g}_{\text{at altitude } y}\right| = G\frac{m_{\text{Earth}}}{(R+y)^2}$$

$$\left|\vec{g}_{\text{at altitude } y}\right| = 0.99\left|\vec{g}_{\text{at altitude surface}}\right|$$

$$G\frac{m_{\text{Earth}}}{(R+y)^2} = 0.99 G\frac{m_{\text{Earth}}}{R^2}$$

$$\cancel{G}\frac{\cancel{m_{\text{Earth}}}}{(R+y)^2} = 0.99\cancel{G}\frac{\cancel{m_{\text{Earth}}}}{R^2}$$

$$\frac{1}{(R+y)^2} = 0.99\frac{1}{R^2}$$

$$(R+y)^2 = (1/0.99)\,R^2$$

$$R^2 + y^2 + 2Ry = (1/0.99)\,R^2$$

$$y^2 + 2Ry + R^2 - (1/0.99)\,R^2 = 0$$

$$y^2 + 2Ry + R^2\left(1 - (1/0.99)\right) = 0$$

Substitute Earth's radius and solve using the quadratic formula or the `solve` function on your calculator.

$$y^2 + 2\left(6.4 \times 10^6\,\text{m}\right)y + \left(6.4 \times 10^6\,\text{m}\right)^2\left(1 - (1/0.99)\right) = 0$$

$$y^2 + 2\left(6.4 \times 10^6\,\text{m}\right)y + -\left(4.137 \times 10^{11}\,\text{m}\right)^2 = 0$$

$$y = 3.2 \times 10^4\,\text{m}$$

Check the answer by calculating g.

$$\left|\vec{g}_{\text{at altitude } y}\right| = G\frac{m_{\text{Earth}}}{(R+y)^2}$$

$$= \left(6.6738 \times 10^{-11}\,\text{N} \cdot \text{m}^2/\text{kg}^2\right)\frac{6 \times 10^{24}\,\text{kg}}{\left(6.4 \times 10^6\,\text{m} + 32000\,\text{m}\right)^2}$$

$$= 9.7\,\text{N/kg}$$

and $\frac{9.7}{9.8} = 99\%$ as expected.

P29:
  Solution:

(a) Sketch a picture of the star and planet.

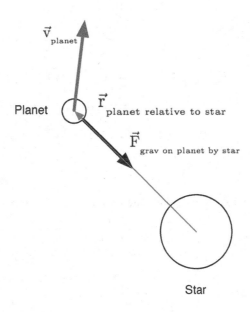

Calculate the net force on the planet, which is the gravitational force by the star.

$$\vec{r} = \vec{r}_{\text{planet}} - \vec{r}_{\text{star}}$$
$$= \left\langle 3 \times 10^{12}, 4 \times 10^{12}, 0 \right\rangle \text{m} - \left\langle 5 \times 10^{12}, 2 \times 10^{12}, 0 \right\rangle \text{m}$$
$$= \left\langle -2 \times 10^{12}, 2 \times 10^{12}, 0 \right\rangle \text{m}$$

$$|\vec{r}| = \sqrt{\left(-2 \times 10^{12}\right)^2 + \left(2 \times 10^{12}\right)^2 + (0)^2} \, \text{m}$$
$$= 2.83 \times 10^{12} \, \text{m}$$

$$\hat{r} = \frac{\vec{r}}{|\vec{r}|}$$
$$= \left\langle -0.707, 0.707, 0 \right\rangle$$

$$\left| \vec{F}_{\text{grav on planet by star}} \right| = G \frac{m_{\text{planet}} m_{\text{star}}}{\left| \vec{r} \right|^2}$$

$$= \frac{\left(6.6738 \times 10^{-11} \text{N} \cdot \text{m}^2/\text{kg}^2 \right) \left(3 \times 10^{24} \text{kg} \right) \left(7 \times 10^{30} \text{kg} \right)}{\left(2.83 \times 10^{12} \text{m} \right)^2}$$

$$= 1.75 \times 10^{20} \text{N}$$

$$\vec{F}_{\text{grav on planet by star}} = \left| \vec{F}_{\text{grav}} \right| (-\hat{r})$$

$$= \left(1.75 \times 10^{20} \text{N} \right) \left(-\langle -0.707, 0.707, 0 \rangle \right)$$

$$= \left\langle 1.24 \times 10^{20}, -1.24 \times 10^{20}, 0 \right\rangle \text{N}$$

Apply the momentum principle to calculate the momentum of the planet after 1×10^6 s. Define the system to be the planet.

$$\vec{p}_{\text{f}} = \vec{p}_{\text{i}} + \vec{F}_{\text{net}} \Delta t$$

$$= m \vec{v}_{\text{i}} + \vec{F}_{\text{net}} \Delta t$$

$$= \left(3 \times 10^{24} \text{kg} \right) \left(\left\langle 0.3 \times 10^4, 1.5 \times 10^4, 0 \right\rangle \text{m/s} \right) + \left(\left\langle 1.24 \times 10^{20}, -1.24 \times 10^{20}, 0 \right\rangle \text{N} \right) \left(1 \times 10^6 \text{s} \right)$$

$$= \left\langle 9.12 \times 10^{27}, 4.49 \times 10^{28}, 0 \right\rangle \text{kg} \cdot \text{m/s}$$

Solve for the final velocity of the planet.

$$\vec{v}_{\text{f}} = \frac{\vec{p}_{\text{f}}}{m}$$

$$= \frac{\left\langle 9.12 \times 10^{27}, 4.49 \times 10^{28}, 0 \right\rangle \text{kg} \cdot \text{m/s}}{3 \times 10^{24} \text{kg}}$$

$$= \left\langle 3.04 \times 10^3, 1.496 \times 10^4, 0 \right\rangle \text{m/s}$$

(b) The position of the planet is found by using the definition of average velocity.

$$\vec{r}_{\text{f}} = \vec{r}_{\text{i}} + \vec{v}_{\text{avg}} \Delta t$$

Use the approximation $\vec{v}_{\text{avg}} \approx \vec{v}_{\text{f}}$.

$$\vec{r}_{\mathrm{f}} = \left\langle 3.004 \times 10^{12}, 4 \times 10^{12}, 0 \right\rangle \mathrm{m} + \left(\left\langle 3.04 \times 10^{3}, 1.496 \times 10^{4}, 0 \right\rangle \mathrm{m/s} \right) \left(1 \times 10^{6} \, \mathrm{s} \right)$$

$$= \left\langle 3.004 \times 10^{12}, 4.01 \times 10^{12}, 0 \right\rangle \mathrm{m}$$

(c) The final velocity and final position are both approximate since the net force on the planet is not constant and since the final velocity is used in place of the average velocity. These approximations improve in the limit as $\Delta t \to 0$. Thus, the approximations would be worse for larger Δt like $1 \times 10^{9} \, \mathrm{s}$.

P33:
Solution:

If the positively charged particle were at the midpoint of the line joining the two negatively charged particles, the net force on it would be zero (a symmetry argument). But the symmetry is broken, and the positively charged particle is closer to the negatively charged particle on the right hand side. Thus, the interaction between the positively charged particle and the right hand negatively charged particle dominated the net force on the positively charged particle. Therefore, the net force will be toward the right. Note that no "formula" is necessary to answer this question. The solution is in the geometry.

P37:
Solution:

(a)
$$\vec{r}_{21} = \vec{r}_{2} - \vec{r}_{1} = \langle -0.4, 0.4, 0 \rangle \, \mathrm{m} - \langle 0.5, -0.2, 0 \rangle \, \mathrm{m}$$
$$\vec{r}_{21} \approx \langle -0.9, 0.6, 0 \rangle \, \mathrm{m}$$

(b)
$$|\vec{r}_{21}| = \sqrt{r_{x,21}^{2} + r_{y,21}^{2} + r_{z,21}^{2}} \approx \sqrt{(-0.9)^{2} + (0.6)^{2} + (0)^{2}} \, \mathrm{m}$$
$$|\vec{r}_{21}| \approx 1.08 \, \mathrm{m}$$

(c)
$$\hat{r}_{21} = \frac{\vec{r}_{21}}{|\vec{r}_{21}|} \approx \frac{\langle -0.9, 0.6, 0 \rangle \, \mathrm{m}}{1.08 \, \mathrm{m}} \approx \langle -0.833, 0.556, 0 \rangle$$

(d)
$$\left| \vec{F}_{\mathrm{grav},21} \right| = G \frac{m_{1} m_{2}}{|\vec{r}_{21}|^{2}}$$

$$\left| \vec{F}_{\mathrm{grav},21} \right| \approx \left(6.6738 \times 10^{-11} \mathrm{N} \cdot \mathrm{m}^{2}/\mathrm{kg}^{2} \right) \left(\frac{\left(2 \times 10^{-3} \mathrm{kg} \right)^{2}}{(1.08 \, \mathrm{m})^{2}} \right) \approx 2.29 \times 10^{-16} \mathrm{N}$$

(e)
$$\vec{F}_{\mathrm{grav},21} = \left| \vec{F}_{\mathrm{grav},21} \right| \hat{F}_{\mathrm{grav},21} = \left| \vec{F}_{\mathrm{grav},21} \right| (-\hat{r}_{21})$$
$$\vec{F}_{\mathrm{grav},21} \approx \left(2.29 \times 10^{-16} \mathrm{N} \right) \langle 0.833, -0.556, 0 \rangle \approx \left\langle 1.91 \times 10^{-16}, -1.27 \times 10^{-16}, 0 \right\rangle \mathrm{N}$$

Make sure this result indicates an attractive interaction and agrees with your diagram.

(f)

$$\left|\vec{F}_{\text{elec},21}\right| = \frac{1}{4\pi\epsilon_{\text{o}}} \frac{\left|q_1 q_2\right|}{\left|\vec{r}_{21}\right|^2}$$

$$\left|\vec{F}_{\text{elec},21}\right| \approx \left(8.9876 \times 10^{9}\,\text{N} \cdot \text{m}^2/\text{C}^2\right) \left(\frac{\left|\left(-2 \times 10^{-9}\,\text{C}\right)\left(-4 \times 10^{-9}\,\text{C}\right)\right|}{(1.08\,\text{m})^2}\right) \approx 6.17 \times 10^{-8}\,\text{N}$$

(g)

$$\vec{F}_{\text{elec},21} = \left|\vec{F}_{\text{elec},21}\right| \hat{F}_{\text{elec},21} = \left|\vec{F}_{\text{elec},21}\right| \left(\hat{r}_{21}\right)$$

$$\vec{F}_{\text{elec},21} \approx \left(6.17 \times 10^{-8}\,\text{N}\right) \langle -0.833, 0.556, 0 \rangle \approx \left\langle -5.14 \times 10^{-8}, 3.43 \times 10^{-8}, 0 \right\rangle \text{N}$$

Make sure this result indicates a repulsive interaction and agrees with your diagram.

(h)

$$\frac{\left|\vec{F}_{\text{elec},21}\right|}{\left|\vec{F}_{\text{grav},21}\right|} \approx \frac{6.17 \times 10^{-8}\,\text{N}}{2.29 \times 10^{-16}\,\text{N}} \approx 2.69 \times 10^{8}$$

(i) Both gravitational and electric interactions vary with distance squared, but the **ratio** of these two interactions is independent of distance for a given pair of interacting particles. Therefore, the ratio of the magnitude of the electric force to the magnitude of the gravitational force is still **2.69e8**.

P41:
Solution:

(a) The total initial momentum of the system is the sum of the momenta of the two balls. Name them A and B.

$$\begin{aligned}
\vec{p}_{\text{sys},i} &= \vec{p}_{\text{A},i} + \vec{p}_{\text{B},i} \\
&= m_{\text{A}} \vec{v}_{\text{A},i} + m_{\text{B}} \vec{v}_{\text{B},i} \\
&= 0.3\,\text{kg} \langle 4, -3, 2 \rangle\,\text{m/s} + 0.5\,\text{kg} \langle 2, 1, 4 \rangle\,\text{m/s} \\
&= \langle 1.2, -0.9, 0.6 \rangle\,\text{kg} \cdot \text{m/s} + \langle 1, 0.5, 2 \rangle\,\text{kg} \cdot \text{m/s} \\
&= \langle 2.2, -0.4, 2.6 \rangle\,\text{kg} \cdot \text{m/s}
\end{aligned}$$

(b) The system is near Earth, so the gravitational force on the system is

$$\begin{aligned}
\vec{F}_{\text{grav by Earth on sys}} &= m\vec{g} \\
&= \langle 0, -mg, 0 \rangle \\
&= \langle 0, -\left(0.3\,\text{kg} + 0.5\,\text{kg}\right)\left(9.80\,\text{m/s}^2\right), 0 \rangle \\
&= \langle 0, -7.84, 0 \rangle\,\text{N}
\end{aligned}$$

(c) Use the momentum principle

$$\Delta \vec{p}_{\text{sys}} = \langle 2.2, -0.4, 2.6 \rangle \, \text{kg} \cdot \text{m/s} + (\langle 0, -7.84, 0 \rangle \, \text{N}) \, (0.1 \, \text{s})$$
$$= \langle 2.2, -0.4, 2.6 \rangle \, \text{kg} \cdot \text{m/s} + \langle 0, -0.784, 0 \rangle \, \text{kg} \cdot \text{m/s}$$
$$= \langle 2.2, -1.2, 2.6 \rangle \, \text{kg} \cdot \text{m/s}$$

Note that the x and z momenta of the system are constant (i.e. they did not change) since the net force on the system is only in the y-direction.

P43:

Solution:

First, sketch a picture. The CM is at the origin and the position of Earth is $\vec{r}_E$ and the position of the Moon is $\vec{r}_M$. The total distance between Earth and Moon is $r = |\vec{r}_E| + |\vec{r}_M|$.

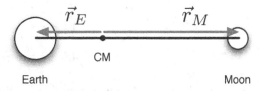

$$M_E = 5.9736 \times 10^{24} \, \text{kg}$$
$$M_M = 7.3459 \times 10^{22} \, \text{kg}$$
$$r = |\vec{r}_E| + |\vec{r}_M| = 4 \times 10^8 \, \text{m}$$
$$\vec{r}_{\text{CM}} = \langle 0, 0, 0 \rangle$$

Starting with the equation for the center of mass position for a two-body system gives:

$$\vec{r}_{\text{CM}} = \frac{M_E \vec{r}_E + M_M \vec{r}_M}{M_E + M_M}$$
$$\langle 0, 0, 0 \rangle = \frac{M_E \vec{r}_E + M_M \vec{r}_M}{M_E + M_M}$$
$$M_E \vec{r}_E = -M_M \vec{r}_M$$
$$M_E |\vec{r}_E| = M_M |\vec{r}_M|$$

Now we have two equations and two unknowns:

$$\text{Eq 1:} \quad |\vec{r}_E| + |\vec{r}_M| = 4 \times 10^8 \, \text{m}$$
$$\text{Eq 2:} \quad M_E |\vec{r}_E| = M_M |\vec{r}_M|$$

Solve for $|\vec{r}_M|$ in the second equation.

$$M_E |\vec{r}_E| = M_M |\vec{r}_M|$$
$$|\vec{r}_M| = \frac{M_E |\vec{r}_E|}{M_M}$$

Substitute this expression into the first equation to solve for the distance from the CM to Earth.

$$|\vec{r}_E| + |\vec{r}_M| = 4 \times 10^8 \, m$$
$$|\vec{r}_E| + \frac{M_E |\vec{r}_E|}{M_M} = 4 \times 10^8 \, m$$
$$|\vec{r}_E| \left(1 + \frac{M_E}{M_M}\right) = 4 \times 10^8 \, m$$
$$|\vec{r}_E| = \frac{4 \times 10^8 \, m}{\left(1 + \frac{M_E}{M_M}\right)}$$
$$= 4.6 \times 10^6 \, m$$

Earth's radius is 6.4×10^6 m, so the CM of the Earth-Moon system is actually inside of Earth's radius (unlike what is shown in our initial picture).

P53:
Solution:

(a)

$$\vec{p}_{sys,i} = \vec{p}_{A,i} + \vec{p}_{B,i}$$
$$= \langle 20, -5, 0 \rangle \, kg \cdot m/s + \langle 5, 6, 0 \rangle \, kg \cdot m/s$$
$$= \langle 25, 1, 0 \rangle \, kg \cdot m/s$$

(b) The impulse on the system of the two objects is due to the net external force on the system. During the small time interval of the collision, the net force on the system is approximately zero, so the impulse on the system is approximately zero.

(c)

$$\vec{p}_{sys,f} = \vec{p}_{A,i} + \cancelto{0}{\vec{F}_{net} \Delta t}$$
$$\vec{p}_{sys,f} = \vec{p}_{sys,i}$$
$$= \langle 25, 1, 0 \rangle \, kg \cdot m/s$$

(d)

$$\vec{p}_{\text{sys,f}} = \vec{p}_{\text{A,f}} + \vec{p}_{\text{B,f}}$$
$$\vec{p}_{\text{B,f}} = \vec{p}_{\text{sys,f}} - \vec{p}_{\text{A,f}}$$
$$= \langle 25, 1, 0 \rangle \, \text{kg} \cdot \text{m/s} - \langle 18, 5, 0 \rangle \, \text{kg} \cdot \text{m/s}$$
$$= \langle 7, -4, 0 \rangle \, \text{kg} \cdot \text{m/s}$$

P57:

Solution:

Define the system as the two rocks. The net external force on the system is zero. Apply the momentum principle.

$$\vec{p}_{\text{sys,f}} = \vec{p}_{\text{sys,i}} + \cancelto{0}{\vec{F}_{\text{net}}} \Delta t$$
$$\vec{p}_{\text{sys,f}} = \vec{p}_{\text{sys,i}}$$

The momentum of the system is the sum of the momenta of the two rocks. The final velocity of each rock is the same since they stick together.

$$\vec{p}_{\text{A,f}} + \vec{p}_{\text{B,f}} = \vec{p}_{\text{A,i}} + \vec{p}_{\text{B,i}}$$
$$m_{\text{A}} \vec{v}_{\text{f}} + m_{\text{B}} \vec{v}_{\text{f}} = m_{\text{A}} \vec{v}_{\text{A,i}} + m_{\text{B}} \vec{v}_{\text{B,i}}$$
$$\left(m_{\text{A}} + m_{\text{B}} \right) \vec{v}_{\text{f}} = m_{\text{A}} \vec{v}_{\text{A,i}} + m_{\text{B}} \vec{v}_{\text{B,i}}$$
$$\vec{v}_{\text{f}} = \frac{m_{\text{A}} \vec{v}_{\text{A,i}} + m_{\text{B}} \vec{v}_{\text{B,i}}}{m_{\text{A}} + m_{\text{B}}}$$
$$= \frac{(15\,\text{kg}) \left(\langle 10, -30, 0 \rangle \, \text{m/s} \right) + (32\,\text{kg}) \left(\langle 15, 12, 0 \rangle \, \text{m/s} \right)}{15\,\text{kg} + 32\,\text{kg}}$$
$$= \langle 13.4, -1.4, 0 \rangle \, \text{m/s}$$

P61:

Solution:

Sketch a picture of the system before and after the package is launched, as shown in the figure below.

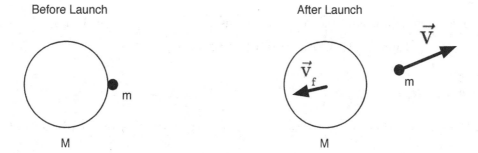

Treat the objects as point particles. Define the system to be the space station and the package. Assume that $m << M$ so that the center of mass of the system before the package is launched is very close to the center of mass of the space station. With this assumption, the initial momentum of the system (before the launch) is zero. Also, assume that the net external force on the system is zero. Apply the momentum principle.

$$\vec{p}_{sys,f} = \vec{p}_{sys,i} + \cancel{\vec{F}_{net}}^{0}\Delta t$$
$$\vec{p}_{station,f} + \vec{p}_{package,f} = \vec{p}_{station,i} + \vec{p}_{package,i}$$
$$= 0 + 0$$

Thus,

$$\vec{p}_{station,f} = -\vec{p}_{package,f}$$
$$M\vec{v}_f = -m\vec{v}$$
$$\vec{v}_f = -\left(\frac{m}{M}\right)\vec{v}$$
$$\vec{v}_f = -\left(\frac{m}{M}\right)\langle v\cos\theta, v\sin\theta, 0\rangle$$

Written in component form:

$$v_{fx} = -\left(\frac{m}{M}\right)v\cos\theta$$

and

$$v_{fy} = -\left(\frac{m}{M}\right)v\sin\theta$$

As a result of launching the package the space station recoils in the opposite direction with equal magnitude and opposite momentum as the package.

CP65:
Solution:

(a) Here's a sample program that uses the initial conditions given in the question. It produces an elliptical orbit.

```
from __future__ import division, print_function
from visual import *

RE = 6.4e6 #radius of Earth

spacecraft = sphere(pos=(-10*RE, 0, 0), color=color.cyan, radius=0.25*RE, make_trail
    =true)
```

```
Earth = sphere (color=color.blue, radius=RE)

m=1.5e4 #mass of spacecraft
ME = 6e24
G = 6.67e-11

v=vector(0,3e3,0) #initial velocity of spacecraft
p=m*v #initial momentum of spacecraft

t=0
dt=60 #time step

rmag=mag(spacecraft.pos); #distance of spacecraft from Earth

while rmag>RE: #stop if rmag < RE
    rate(1000)
    r=spacecraft.pos
    rmag=mag(r)
    runit=r/rmag

    Fgrav=-G*m*ME/rmag**2*runit
    Fnet=Fgrav

    p = p + Fnet*dt
    v = p/m
    spacecraft.pos = spacecraft.pos + v*dt

    t = t + dt
```

The output is shown below.

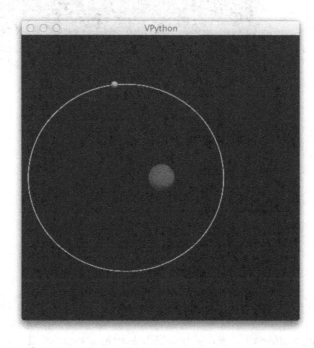

By increasing the speed (to 4×10^3 m/s for example), you can get the spaceship to leave Earth and not return. An example of this "open orbit" is shown below.

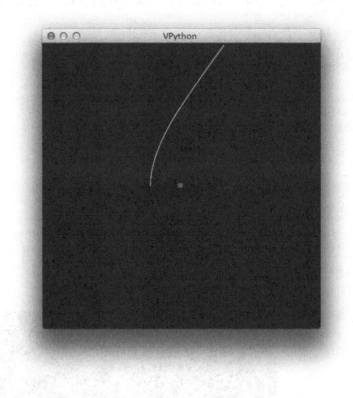

(b) A speed of 2×10^3 m/s produces an elliptical orbit with Earth at a focus of the ellipse. Increasing the speed to 3×10^3 m/s produces an elliptical orbit with Earth at the other focus of the ellipse. (An ellipse has two foci. So evidently the initial speed affects the focus at which Earth is located, if the orbit is elliptical.) The output for a speed of 3×10^3 m/s is shown below.

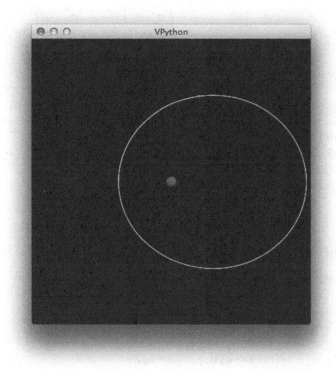

(c) To add arrows, you have to first get an idea of the scale factor you will use to make the arrows visible. Thus, add a line inside the `while` loop to print the magnitude of the net force and the magnitude of the momentum.

print(mag(Fnet), mag(p))

For a speed of 3×10^3 m/s, the print statement gives an initial net force and initial momentum of 1466 N and 4.5×10^7 kg · m/s.

Now create the arrows and create scale factors. Use the initial distance of the spacecraft from Earth to get a reasonable estimate for the size of the scale factor. I used

Fscale=12*RE/1466
Pscale=12*RE/4.5e7

Here is a final program that displays arrows for the net force and the momentum.

```
from __future__ import division, print_function
from visual import *

RE = 6.4e6 #radius of Earth

spacecraft = sphere(pos=(-10*RE, 0, 0), color=color.cyan, radius=0.25*RE, make_trail
    =true)
Earth = sphere (color=color.blue, radius=RE)

m=1.5e4 #mass of spacecraft
ME = 6e24
G = 6.67e-11

v=vector(0,2.5e3,0) #initial velocity of spacecraft
p=m*v #initial momentum of spacecraft
```

```
t=0
dt=60 #time step

rmag=mag(spacecraft.pos); #distance of spacecraft from Earth

Parrow=arrow(pos=spacecraft.pos, axis=(0,0,0), color=color.yellow)
Farrow=arrow(pos=spacecraft.pos, axis=(0,0,0), color=color.magenta)
Fscale=12*RE/1466
Pscale=12*RE/4.5e7

while rmag>RE: #stop if rmag < RE
    rate(500)
    r=spacecraft.pos
    rmag=mag(r)
    runit=r/rmag

    Fgrav=-G*m*ME/rmag**2*runit
    Fnet=Fgrav

    p = p + Fnet*dt
    v = p/m
    spacecraft.pos = spacecraft.pos + v*dt

#    print(mag(Fnet), mag(p))
    Farrow.pos=spacecraft.pos
    Farrow.axis=Fscale*Fnet
    Parrow.pos=spacecraft.pos
    Parrow.axis=Pscale*p

    t = t + dt
```

Here is the output.

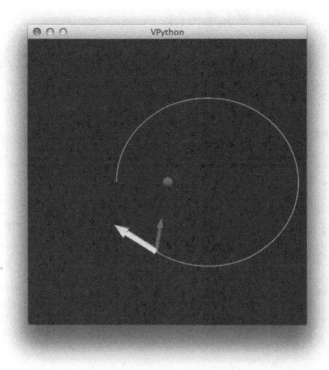

(d) A speed of 2.5×10^3 m/s gives a circular orbit. The output is shown below.

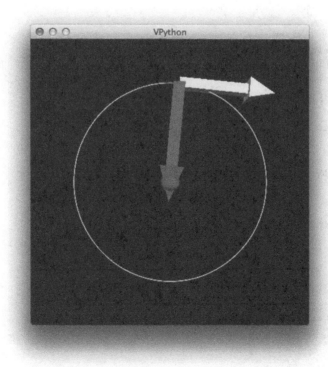

(e) A time step of $\Delta t = 3600$ s (1 hour) is approximately the maximum time step that gives a nearly closed orbit for at least a few revolutions.

CP71:
  Solution:

(a) Modify the program in 66. A step size of 10 s can be used. Here's an example program where Ranger's initial velocity is directly toward Moon. Note that the radii of the spheres are larger than the actual radii of Earth and Moon, in order to make them visible in the simulation. A print statement is added after the while loop (outside the while loop) so that the clock reading can be printed after Ranger crashes into either Earth or Moon.

```
from __future__ import division, print_function
from visual import *

RE = 6.4e6 #radius of Earth
RM = 1.75e6 #radius of Moon
h = 5e4 #initial altitude of Ranger

Earth = sphere(pos=(0,0,0), color=color.blue, radius=5*RE)
Moon = sphere(pos=(4e8,0,0), color=color.white, radius=0.5*Earth.radius)
ranger = sphere(pos=(RE+h, 0, 0), color=color.cyan, radius=0.25*Earth.radius)

m=173 #mass of ranger
ME = 6e24 #mass of Earth
MM = 7e22 #mass of Moon
G = 6.67e-11

v=vector(1.11e4,0, 0) #initial velocity of ranger
p=m*v #initial momentum of ranger

t=0
dt=10

rmag=mag(ranger.pos); #distance of ranger from Earth
rrelmoonmag=mag(ranger.pos-Moon.pos) #distance of ranger from Moon

trail=curve(color=ranger.color)

while rmag>RE and rrelmoonmag>RM: #stop if rmag < RE or rrelmoonmag < RM
    rate(1000)

    #calculate Fgrav on ranger by Earth
    r=ranger.pos
    rmag=mag(r)
    runit=r/rmag
    FgravE=-G*m*ME/rmag**2*runit

    #calculate Fgrav on ranger by Moon
    rrelmoon=ranger.pos - Moon.pos
    rrelmoonmag=mag(rrelmoon)
    rrelmoonunit=rrelmoon/rrelmoonmag
    FgravM=-G*m*MM/rrelmoonmag**2*rrelmoonunit

    #calculate net force
    Fnet=FgravE + FgravM

    #update momentum and position
```

```
    p = p + Fnet*dt
    v = p/m
    ranger.pos = ranger.pos + v*dt

    trail.append(pos=ranger.pos)

    t = t + dt

print "crashed at t=",t
```

(b) The minimum initial speed of the Ranger to reach Moon is 1.2×10^4 m/s (rounded to two significant figures). With initial speeds less than this speed, the Ranger reaches a turning point (zero speed) and returns to Earth. To three significant figures, with the constants defined as in the sample program, the initial speed was found to be 1.11×10^4 m/s.

(c) Use an initial speed of $(1.1)\left(1.2 \times 10^4 \text{ m/s}\right) = 1.32 \times 10^4$ m/s. Use the print statement after the while loop to print the clock reading in seconds. Convert this to hours and days. The duration of the trip is about 15 hours, or 0.6 days.

(d) After the while loop, use print "speed = ", mag(v) to print the magnitude of Ranger's velocity. With an initial speed of 1.32×10^4 m/s, the impact speed at Moon is 7500 m/s.

4   Chapter 4: Contact Interactions

Q03:
Solution:

(a) Neglect the mass of the rope and assume that tension is uniform throughout the rope. Apply the momentum principle to the climber. Sketch a free-body diagram. Define the system to be the climber.

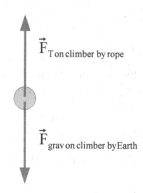

$$\vec{F}_{\text{net}} = \frac{\Delta \vec{p}}{\Delta t}$$

The climber's momentum is constant (since the climber is "motionless"), so

$$\vec{F}_{\text{net}} = 0$$

Sum the forces from the free-body diagram.

$$\vec{F}_{\text{T by rope}} + \vec{F}_{\text{grav by Earth}} = 0$$
$$\vec{F}_{\text{T by rope}} = -\vec{F}_{\text{grav by Earth}}$$
$$= -\left\langle 0, -mg, 0 \right\rangle$$
$$= \left\langle 0, mg, 0 \right\rangle$$
$$= \left\langle 0, (55\,\text{kg})\left(9.80\,\text{m/s}^2\right), 0 \right\rangle$$
$$= \left\langle 0, 539, 0 \right\rangle \text{N}$$

$$\left| \vec{F}_{\text{T by rope}} \right| = 539\,\text{N}$$

(b)

$$m = 88\,\text{kg}$$
$$\vec{F}_{\text{T by rope}} = \left\langle 0, mg, 0 \right\rangle$$
$$= \left\langle 0, (88\,\text{kg})\left(9.80\,\text{m/s}^2\right), 0 \right\rangle$$
$$= \left\langle 0, 862, 0 \right\rangle \text{N}$$

(c) Both (2) and (3) are true. Model the rope as balls connected by springs in one dimension. Tension (i.e. a force applied to the rope) causes the interatomic springs (i.e. bonds) to stretch. As a result the "atoms" in the one-dimensional model get further apart.

Q09:
Solution:

(b) $Y_A = Y_B$ because both wires are made of pure copper.

Q15:
Solution:

(a)

$$T = 2\pi\sqrt{\frac{m}{k}}$$
$$T \propto \sqrt{m}$$

If m is doubled, T increases by a factor $\sqrt{2}$. So if $T = 1\,\text{s}$, doubling the mass gives a period of $\sqrt{2}\,(1\,\text{s}) = 1.4\,\text{s}$

(b)

$$T = 2\pi\sqrt{\frac{m}{k}}$$
$$T \propto \frac{1}{\sqrt{k}}$$

Doubling the stiffness causes T to change by a factor of $\frac{1}{\sqrt{2}}$. Thus, if $T = 1\,\text{s}$, then doubling the stiffness results in a period of $0.71\,\text{s}$.

(c)

$$\frac{1}{k_{\text{eff}}} = N\frac{1}{k}$$
$$k = Nk_{\text{eff}}$$
$$= 2k_{\text{eff}}$$

Thus, cutting the spring in half doubles the stiffness. Since

$$T \propto \frac{1}{\sqrt{k}}$$

doubling the stiffness changes T by a factor of $\frac{1}{\sqrt{2}} = 0.71$. Thus, a period of $1\,\text{s}$ becomes a period of $(0.71)\,(1\,\text{s}) = 0.71\,\text{s}$.

(d) Period is independent of amplitude. Therefore, the period will remain $1\,\text{s}$.

(e) Period is independent of g. Therefore, the period will remain $1\,\text{s}$.

P21:

   Solution:

$$
\begin{aligned}
d &= 1.7 \times 10^{-10}\,\text{m} \\
m &= 8.2 \times 10^{-26}\,\text{kg} \\
\rho &= ?
\end{aligned}
$$

The interatomic distance is the diameter of an atom. The volume of space associated with each atom is a cube of length d and volume d^3, as shown below.

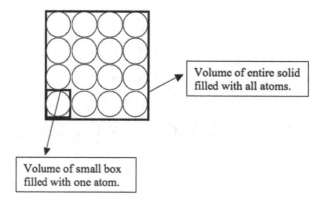

The density of the material is

$$
\begin{aligned}
\rho &= \frac{\text{mass of all atoms}}{\text{volume of material}} = \frac{\text{mass of 1 atom}}{d^3} \\
\rho &= \frac{8.2 \times 10^{-26}\,\text{kg}}{\left(1.7 \times 10^{-10}\,\text{m}\right)^3} \\
&= 1.7 \times 10^{4}\,\text{kg/m}^3
\end{aligned}
$$

P25:

   Solution:

Springs in series have an effective stiffness

$$
\frac{1}{k_{s,\text{eff}}} = \frac{1}{k_{s,1}} + \frac{1}{k_{s,2}} + \dots
$$

For N identical springs,

$$
\frac{1}{k_{s,\text{eff}}} = N\frac{1}{k_{s}}
$$

Each individual spring has a stiffness

$$k_s = N k_{s,\text{eff}}$$
$$= (50)\,(270\,\text{N/m})$$
$$= 1.35 \times 10^4\,\text{N/m}$$

P31:
Solution:

Start by calculating Young's modulus for copper.

$$Y = \frac{\left|\vec{F}\right|/A}{\Delta L/L}$$

$$Y = \frac{(36\,\text{kg})\,(9.80\,\text{m/s}^2)\,/\left((\pi)\left(0.7 \times 10^{-3}\,\text{m}\right)^2\right)}{(0.00183\,\text{m})\,/\,(0.95\,\text{m})}$$

$$Y \approx 1.90 \times 10^{11}\,\text{N/m}^2$$

Next, calculate the mass of one copper atom.

$$m_{\text{Cu}} = \frac{63\,\text{g/mol}}{6.0221 \times 10^{23}\,\text{mol}^{-1}}$$

$$m_{\text{Cu}} \approx 1.05 \times 10^{-22}\,\text{kgg}$$

Now, use the density (ρ) and atom's mass to calculate an approximate interatomic spacing, assuming a cubic atom.

$$d \approx \sqrt[3]{\frac{m_{\text{Cu}}}{\rho}}$$

$$d \approx \sqrt[3]{\frac{1.05 \times 10^{-22}\,\text{kgg}}{9\,\text{g/cm}^3}} \approx 2.27 \times 10^{-8}\,\text{cm} \approx 2.27 \times 10^{-10}\,\text{m}$$

Finally, use the Young's modulus and interatomic spacing to calculate the interatomic stiffness.

$$k_s \approx Y d$$

$$k_s \approx \left(1.2 \times 10^{11}\,\text{N/m}^2\right)\left(2.27 \times 10^{-10}\,\text{m}\right) \approx 27\,\text{N/m}$$

P37:
Solution:

This is a straightforward application of the basic definition of Young's modulus.

$$Y = \frac{\left|\vec{F}\right|/A}{L/\Delta L}$$

$$\Delta L = \frac{\left|\vec{F}\right|/A}{Y/L}$$

$$\Delta L = \frac{\left|\vec{F}\right|/\left(\pi r^2\right)}{Y/L}$$

$$\Delta L \approx \frac{(10\,\text{kg})\left(9.80\,\text{m/s}^2\right)/\left(\pi\left(1.5\times 10^{-3}\,\text{m}\right)^2\right)}{\left(2\times 10^{11}\text{N/m}^2\right)/(3\,\text{m})}$$

$$\Delta L \approx 2.1\times 10^{-4}\,\text{m} \approx 0.21\,\text{mm}$$

P45:

Solution:

Assume a horizontal floor.

Define the system to be the box. Draw a free-body diagram like the one shown below.

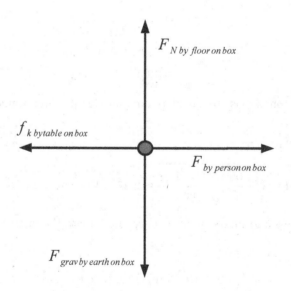

Apply the Momentum Principle

$$\vec{F}_{\text{net}} = \frac{d\vec{p}}{dt}$$

Write it in component form, starting with the y-direction.

$$F_{net,\,y} = \frac{\Delta p_y}{\Delta t}$$
$$= 0$$
$$F_{N \text{ by floor on box},\,y} + F_{grav,\,y} = 0$$
$$F_N + -mg = 0$$
$$F_N = mg$$
$$= (20\,\text{kg})\,(9.80\,\text{m/s}^2)$$
$$= 196\,\text{N}$$

In the x-direction,

$$F_{net,\,x} = \frac{\Delta p_x}{\Delta t}$$
$$F_{person} + f_{k,x} = \frac{\Delta p_x}{\Delta t}$$
$$F_{person} + -\mu_k F_N = \frac{\Delta p_x}{\Delta t}$$
$$90\,\text{N} - (0.25)\,(196\,\text{N}) = \frac{p_{fx} - p_{ix}}{\Delta t}$$
$$90\,\text{N} - 49\,\text{N} = \frac{p_{fx} - p_{ix}}{\Delta t}$$
$$41\,\text{N} = \frac{p_{fx} - p_{ix}}{\Delta t}$$

Solve for v_{fx}.

$$41\,\text{N} = \frac{mv_{fx} - mv_{ix}}{\Delta t}$$
$$41\,\text{N} = \frac{(20\,\text{kg})\,\left(v_{fx} - 3\,\text{m/s}\right)}{0.6\,\text{s}}$$
$$v_{fx} = 3\,\text{m/s} + \left(\frac{41\,\text{N}}{20\,\text{kg}}\right)(0.6)$$
$$= 3\,\text{m/s} + 1.23\,\text{m/s}$$
$$= 4.23\,\text{m/s}$$

To get x_f, use $v_{avg,\,x}$.

$$v_{avg,\,x} = \frac{v_{ix} + v_{fx}}{2}$$
$$= \frac{3\,\text{m/s} + 4.23\,\text{m/s}}{2}$$
$$= 3.62\,\text{m/s}$$

$$x_f = x_i + v_{avg,\,x} \Delta t$$
$$= 8\,\text{m} + (3.62\,\text{m/s})\,(0.6\,\text{s})$$
$$= 10.2\,\text{m}$$

P51:
  Solution:

$$\omega = \sqrt{\frac{k_s}{m}}$$
$$= \sqrt{\frac{8\,\text{N/m}}{2.2\,\text{kg}}}$$
$$= 1.91\,\text{rad/s}$$

$$x = A\cos\left(\omega t + \phi\right)$$

In this case, since $x = +A$ at $t = 0$, $\phi = 0$. So, at $t = 1.15\,\text{s}$,

$$x = A\cos\omega t$$
$$= (0.18\,\text{m})\cos\left((1.91\,\text{rad/s})\,(1.15\,\text{s})\right)$$
$$= -0.105\,\text{m}$$

P57:
  Solution:

First, calculate the angular frequency of the oscillator.

$$\omega = 2\pi f = \frac{2\pi}{T}$$
$$= \frac{2\pi}{0.4\,\text{s}}$$
$$= 15.7\,\text{rad/s}$$

(a) The position as a function of time of a harmonic oscillator is $x = A\cos\left(\omega t\right)$. Its instantaneous velocity is

$$v_x = \frac{\mathrm{d}x}{\mathrm{d}t}$$
$$= -A\omega \sin\left(\omega t\right)$$

Thus, v_x also oscillates sinusoidally with a maximum $A\omega$. The maximum speed is

$$v_{max} = A\omega$$
$$= (0.06\,\text{m})(15.7\,\text{rad/s})$$
$$= 0.94\,\text{m/s}$$

(b) The instantaneous acceleration is

$$a_x = \frac{dv_x}{dt}$$
$$= -A\omega^2 \cos(\omega t)$$

Thus, a_x also oscillates sinusoidally with a maximum $A\omega^2$. The maximum acceleration is

$$a_{max} = A\omega^2$$
$$= (0.06\,\text{m})(15.7\,\text{rad/s})^2$$
$$= 15\,\text{m/s}^2$$

P63:
Solution:

Apply the Momentum Principle to the ball. Define the system to be the ball. Draw a free-body diagram, as shown in the figure below.

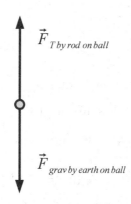

Since the body is in equilibrium,

$$\vec{F}_{\text{net}} = \frac{\Delta \vec{p}}{\Delta t}$$
$$= 0$$
$$\vec{F}_{\text{T}} + \vec{F}_{\text{grav}} = 0$$
$$\vec{F}_{\text{T}} = -\vec{F}_{\text{grav}}$$
$$= -\langle 0, -mg, 0 \rangle$$
$$= \langle 0, mg, 0 \rangle$$
$$= \langle 0, (41\,\text{kg}) \left(9.80\,\text{m/s}^2\right), 0 \rangle$$
$$= \langle 0, 402, 0 \rangle\,\text{N}$$
$$\left| \vec{F}_{\text{T}} \right| = 402\,\text{N}$$

Calculate Young's Modulus,

$$\frac{F_T}{A} = Y \frac{\Delta L}{L}$$
$$Y = \frac{F_T}{A} \frac{L}{\Delta L}$$

The cross-sectional area of the rod is

$$A = (1.5\,\text{mm})(3.1\,\text{mm})$$
$$= \left(1.5 \times 10^{-3}\,\text{m}\right) \left(3.1 \times 10^{-3}\,\text{m}\right)$$
$$= 4.65 \times 10^{-6}\,\text{m}^2$$
$$Y = \left(\frac{402\,\text{N}}{4.65 \times 10^{-6}\,\text{m}^2} \right) \left(\frac{2.6\,\text{m}}{0.002898\,\text{m}} \right)$$
$$= 7.8 \times 10^{10}\,\text{N/m}^2$$

Calculate the diameter of a silver atom. Assume a simple cubic lattice. Find the volume of a cube taken up by a spherical atom.

$$\rho = \left(10.5\,\text{g/cm}^3\right) \left(\frac{1\,\text{kg}}{1000\,\text{g}} \right) \left(\frac{(100\,\text{cm})^3}{1\,\text{m}^3} \right) = 1.05 \times 10^4\,\text{kg/m}^3$$
$$M = 108\,\text{g/mol} = 0.108\,\text{kg/mol}$$

$$V = \left(\frac{1\,\text{m}^3}{1.05 \times 10^4\,\text{kg}} \right) \left(\frac{0.108\,\text{kg}}{1\,\text{mol}} \right) \left(\frac{1\,\text{mol}}{6.02 \times 10^{23}\,\text{atom}} \right)$$

$$= 1.71 \times 10^{-29}\,\text{m}^3$$

$$d = V^{\frac{1}{3}}$$

$$= 2.56 \times 10^{-10}\,\text{m}$$

The interatomic bond stiffness is

$$k_s = Yd$$

$$= \left(7.8 \times 10^{10}\,\text{N/m}^2 \right) \left(2.56 \times 10^{-10}\,\text{m} \right)$$

$$= 20.0\,\text{N/m}$$

Calculate the speed of sound in silver.

$$\text{v} = \sqrt{\frac{k_s}{m_s}}\,d$$

The mass of an atom is

$$m_a = (0.108\,\text{kg/mol}) \left(\frac{1\,\text{mol}}{6.02 \times 10^{23}\,\text{atom}} \right)$$

$$= 1.79 \times 10^{-25}\,\text{kg/atom}$$

$$\text{v} = \sqrt{\frac{20\,\text{N/m}}{1.79 \times 10^{-25}\,\text{kg}}} \left(2.56 \times 10^{10}\,\text{m} \right)$$

$$= 2710\,\text{m/s}$$

P67:
   Solution:

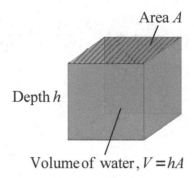

The pressure at depth h is the pressure at the top plus the weight of the volume of water divided by its area.

$$p_{\text{bottom}} = p_{\text{top}} + \frac{mg}{A}$$

Multiply the last term by $\frac{h}{h}$

$$p_{\text{bottom}} = p_{\text{top}} + \frac{mgh}{Ah}$$

Ah is the volume and mass/volume is the density of the water.

$$p_{\text{bottom}} = p_{\text{top}} + \frac{mgh}{V}$$
$$p_{\text{bottom}} = p_{\text{top}} + \rho_{\text{water}}\, gh$$

The density of freshwater is $1000\ \frac{\text{kg}}{\text{m}^3}$. Solve for h.

$$h = \frac{p_{\text{bottom}} - p_{\text{top}}}{\rho_{\text{water}}\, g}$$
$$= \frac{3 \times 10^5\ \frac{\text{N}}{\text{m}^2} - 1 \times 10^5\ \frac{\text{N}}{\text{m}^2}}{\left(1000\ \frac{\text{kg}}{\text{m}^3}\right)\left(9.8\ \frac{\text{N}}{\text{kg}}\right)}$$
$$= 20.4\,\text{m}$$

Saltwater has a greater density than freshwater. It is $1030\ \frac{\text{kg}}{\text{m}^3}$. Thus in seawater,

$$h = \frac{p_{\text{bottom}} - p_{\text{top}}}{\rho_{\text{seawater}}\, g}$$
$$= \frac{3 \times 10^5\ \frac{\text{N}}{\text{m}^2} - 1 \times 10^5\ \frac{\text{N}}{\text{m}^2}}{\left(1030\ \frac{\text{kg}}{\text{m}^3}\right)\left(9.8\ \frac{\text{N}}{\text{kg}}\right)}$$
$$= 19.8\,\text{m}$$

CP71:
   Solution:

(a) First we need to think about the coordinate system. The origin is at the center between the walls. One end of Spring 1 is attached to the left wall which is at the location $\vec{r}_{wall}$. The other end of Spring 1 will be attached to the surface of the ball. Define the vector $\vec{r}_{ball}$ to be the position of the ball. Define the vector $\vec{R}$ to point from the center of the ball to the surface of the ball. The position of the ball relative to the wall is $\vec{r}_{rel} = \vec{r}_{ball} - \vec{r}_{wall}$.

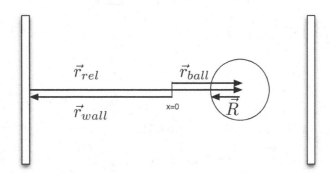

The axis of the spring is $\vec{r}_{rel} - R\hat{r}_{rel}$, where R is the radius of the ball. These vectors can be used to calculate the force by the spring on the ball. The length of the spring is $L = |\vec{r}_{rel}|$. The distance the spring is stretched is $s = L - L_0$. So the force by the spring on the ball is

$$\vec{F}_{spring} = -k_s s \hat{r}_{rel}$$

If $L > L_0$, then s is positive, the spring is stretched, and the force by the spring on the ball is toward the wall. If $L < L_0$, then s is negative, the spring is compressed, and the force by the spring is away from the origin. The example program below uses this equation for the force by the spring to model the motion of the object.

```
from __future__ import division, print_function
from visual import *
from visual.graph import *

m=0.03 #mass in kg
k=4 #spring stiffness in N/m
w=1 #distance to wall from origin
L0=0.5 #relaxed length in m
R=0.05 #radius of the ball in m

leftwall=box(pos=(-w-0.005,0,0), size=(0.01,5*R,5*R), color=color.green)
rightwall=box(pos=(w+0.005,0,0), size=(0.01,5*R,5*R), color=color.green)
ball=sphere(pos=(0,0,0), radius=R, color=color.orange)
spring1=helix(pos=(-w,0,0), axis=ball.pos-vector(-w,0,0), radius=2/3*R, color
    =(0.5,0.5,0.5), thickness=R/5)
spring2=helix(pos=(w,0,0), axis=ball.pos-vector(w,0,0), radius=2/3*R, color
    =(0.5,0.5,0.5), thickness=R/5)

ball.v=vector(6,0,0)
ball.p=m*ball.v
```

```
dt=0.005
t=0

graph=gdisplay(x=430,y=0,width=430, height=450,
                            title='x vs t',
                            xtitle='t (s)',
                            ytitle='y (m)',
                            background=color.black)

function=gcurve(color=color.magenta)


while t<3:
    rate(50)
    #spring1
    rrel=ball.pos-vector(-w,0,0) #position of ball relative to wall
    L=mag(rrel)-R #length of the spring
    s=L-L0 #distance the spring is stretched
    rhat=norm(rrel) #dir of vector pointing to the ball's position
    Fspring1=-k*s*rhat

    #spring2
    rrel=ball.pos-vector(w,0,0) #position of ball relative to wall
    L=mag(rrel)-R #length of the spring
    s=L-L0 #distance the spring is stretched
    rhat=norm(rrel) #dir of vector pointing to the ball's position
    Fspring2=-k*s*rhat

    Fnet=Fspring1+Fspring2
    ball.p=ball.p+Fnet*dt
    ball.pos=ball.pos+ball.p/m*dt

    function.plot(pos=(t,ball.pos.x))


    spring1.axis=ball.pos-vector(-w,0,0)
    spring2.axis=ball.pos-vector(w,0,0)

    t=t+dt
```

The output is shown below.

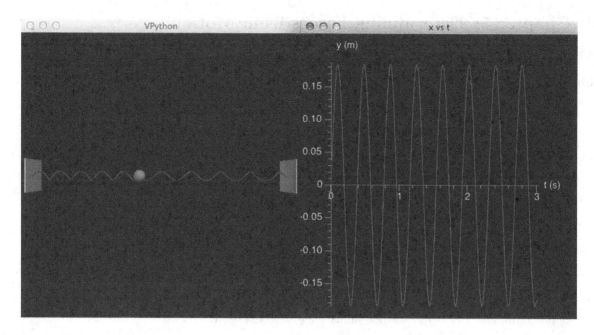

The clock reading of the first peak is $t_1 = 0.0951$ s. The clock reading of the sixth peak (5 cycles after the first peak) is $t_2 = 2.019$ s. The total time for 5 cycles is 1.923 s. The measured period is

$$T = \frac{1.92\,\text{s}}{5\,\text{cycles}} = 0.385\,\text{s}$$

(b) Using $\omega = 2\pi f$, $f = 1/T$, and $\omega = \sqrt{k_{\text{eff}}/m}$, solve for k_{eff} in terms of the period.

$$\omega = \frac{2\pi}{T}$$
$$T = \frac{2\pi}{\omega}$$
$$T = 2\pi\sqrt{\frac{m}{k_{\text{eff}}}}$$
$$k_{\text{eff}} = \frac{4\pi^2 m}{T^2}$$
$$= \frac{4\pi^2 0.03\,\text{kg}}{(0.385\,\text{s})^2}$$
$$= 8.0\,\text{N/m}$$

Thus, $k_{\text{eff}} = 2k_s$. This is exactly what was expected from theory. See P60 for a derivation of k_{eff} from the Momentum Principle.

(c) We can change the amplitude of the oscillation by changing the position of the ball when it is created or the initial speed of the ball. In this case, I chose to change the initial speed of the ball from 3 m/s to 6 m/s. The measured period is the same, 0.385 s. Thus, period is independent of the amplitude, as expected from theory.

5   Chapter 5: Determining Forces from Motion

Q03:
   Solution:

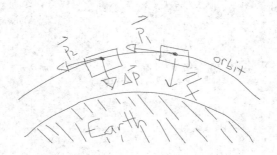

(a) See diagram. The astronaut's momentum is tangent to the orbit.

(b) See diagram. The only force on the astronaut is that of Earth's gravitational attraction, directly toward Earth's center (assuming Earth is spherical). The force must also be perpendicular to the momentum.

(c) See diagram. The momentum is still tangent to the orbit.

(d) The momentum **change** is the vector that is added to the initial momentum to get the final momentum. It is directed perpendicular to the momentum and is therefore parallel to the force. This comes straight from the momentum principle.

(e) The astronaut isn't floating at all. Both the astronaut and the shuttle are in a state of free fall. The astronaut accelerates toward the floor, but the floor accelerates in the same direction at the same rate so the astronaut "misses" the floor.

P11:
   Solution:

Let the left wire be 1, the right wire be 2, and the vertical wire be 3.

(a) $\left|\vec{F}_1\right| \approx 6098$ N, $\left|\vec{F}_2\right| \approx 3980$ N, $\left|\vec{F}_3\right| \approx 7840$ N

(b) For each wire, the strain will just be the stress divided by Young's modulus. Wire 1: $\frac{\Delta L}{L_o} \approx 0.049$, Wire 2: $\frac{\Delta L}{L_o} \approx 0.032$, Wire 3: $\frac{\Delta L}{L_o} \approx 0.063$

P15:
   Solution:

(a) Let the system be both blocks.

$$\begin{aligned}
F_{net,both,x} &= F_x - f_{floor\ on\ 1,x} - f_{floor\ on\ 2,x} = (m_1 + m_2)a_x \\
&= F_x - \mu_{k1}m_1g - \mu_{k2}m_2g \\
&= F_x - (\mu_{k1}m_1 + \mu_{k2}m_2)g \\
a_x &= \frac{F_x - (\mu_{k1}m_1 + \mu_{k2}m_2)g}{m_1 + m_2} \\
a_x &= \frac{(110\ \text{N}) - ((0.3)(12\ \text{kg}) + (0.5)(5\ \text{kg}))(9.8\ \text{m/s}^2)}{17\ \text{kg}} \\
&= 2.95\ \text{m/s}^2
\end{aligned}$$

(b) Now let the system be just block 2.

$$\begin{aligned}
F_{net,2,x} &= F_{1\ on\ 2,x} - f_{floor\ on\ 2,x} = m_2 a_x \\
F_{1\ on\ 2,x} &= m_2 a_x + \mu_2 m_2 g \\
&= \frac{m_2}{m_1 + m_2}\left(F_x - (\mu_1 m_1 + \mu_2 \mu_2)\,g\right) + \mu_2 m_2 g \\
&= \frac{5}{17}(110\ \text{N} - ((0.3)(12\ \text{kg}) + (0.5)(5\ \text{kg}))) + (0.5)(5\ \text{kg})(9.8\ \text{m/s}^2) \\
&= 39.3\ \text{N}
\end{aligned}$$

P21:

Solution:

This problem is basically the same as the previous problem, but you have to remember that the direction of the velocity is the same as the direction of the momentum. You need not explicitly calculate the momentum.

$$\begin{aligned}
\vec{v} &= \langle 4.5 \times 10^4, -1.7 \times 10^4, 0 \rangle\ \text{m/s} \\
\vec{F} &= \langle 1.5 \times 10^{22}, 1.9 \times 10^{23}, 0 \rangle\ \text{N} \\
\vec{F}_\parallel &= \left(\vec{F} \bullet \hat{v}\right)\hat{v} \\
&= \langle -4.97 \times 1022, 1.88 \times 10^{22}, 0 \rangle\ \text{N} \\
\vec{F}_\perp &= \vec{F} - \vec{F}_\parallel \\
&= \langle 6.47 \times 10^{22}, 1.71 \times 10^{23}, 0 \rangle\ \text{N}
\end{aligned}$$

The following VPython script was used for this problem.

```
from __future__ import division, print_function
from visual import *

v = vector(4.5e4,-1.7e4,0)
F = vector(1.5e22,1.9e23,0)
vhat = norm(v)
print(vhat)

Fparallel = dot(F,vhat)*vhat
print (Fparallel)

Fperp = F - Fparallel
```

```
print (Fperp)
```

P25:
Solution:

(a) d

(b) b

(c) Because the proton's speed is constant, $\frac{d|\vec{p}|}{dt}\hat{p} = 0$. Thus,

$$
\begin{aligned}
\frac{d\vec{p}}{dt} &= |\vec{p}|\frac{d\hat{p}}{dt} \\
\left|\frac{d\vec{p}}{dt}\right| &= |\vec{p}|\frac{|\vec{v}|}{R} \\
&= \frac{m|\vec{v}|^2}{R} \text{ since the proton's speed is much less than c.} \\
&= 8.74 \times 10^{-15} \text{ N}
\end{aligned}
$$

P31:
Solution:

Let the system be bucket and water. At the top of the circle, the forces on the system are a downward gravitational force and a downward force from the rope (tension). These must combine to give a net downward centripetal force, toward the circle's center.

$$
\begin{aligned}
\text{F}_{net,y} &= -Mg - \text{T}_y = -\frac{M|\vec{v}|^2}{R} \\
\text{T}_y &= -Mg + \frac{M|\vec{v}|^2}{R} \\
&= \left(-g + \frac{|\vec{v}|^2}{R}\right)M \\
&= \left(-(9.8 \text{ N/kg}) + \left(\frac{(4 \text{ m/s})^2}{1.3 \text{ m}}\right)\right)(3.5 \text{ kg}) \\
&= 8.78 \text{ N}
\end{aligned}
$$

P35:
Solution:

The net force on the roller coaster at any instant is

$$
\vec{F}_{net} = \vec{F}_{track} + \vec{F}_{grav}
$$

Since the roller coaster travels in a circle with constant speed,

$$\left|\vec{F}_{net}\right| = |\vec{p}| \frac{|\vec{v}|}{R}$$
$$= \frac{m |\vec{v}|^2}{R} \text{ since the speed is much less than c.}$$

At the top of the roller coaster, the net force on the roller coaster is toward the center of the circle which is in the downward $(-y)$ direction. The gravitational force on the roller coaster is also in the downward $(-y)$ direction. At the minimum speed needed to make it around the loop, $\vec{F}_{track} = \vec{0}$ at the top of the loop. Thus,

$$\vec{F}_{net} = \vec{F}_{grav}$$
$$\left|\vec{F}_{net}\right| = \left|\vec{F}_{grav}\right|$$
$$\frac{m |\vec{v}|^2}{R} = mg$$
$$|\vec{v}| = \sqrt{Rg}$$

A reasonable R is on the order of 10 m which gives a minimum speed of about 10 m/s.

P39:
Solution:

First, convert period from days to seconds. $T = 1.04 \text{ day} = 8.99 \times 10^4$ s. Define the system to be the NEAR spacecraft. The net force on NEAR is equal to the gravitational force by Eros on NEAR. Note that the speed of NEAR is much less than c. Thus,

$$\left|\vec{F}_{net}\right| = \left|\vec{F}_{grav}\right|$$
$$|\vec{p}| \frac{|\vec{v}|}{R} = \frac{GMm}{R^2}$$
$$\frac{m |\vec{v}|^2}{R} = \frac{GMm}{R^2}$$
$$|\vec{v}|^2 = \frac{MG}{R}$$

The speed of the spacecraft is $v = \frac{2\pi R}{T}$. Substitute this to get an expression that relates the period and radius of the spacecraft's orbit.

$$\left(\frac{2\pi R}{T}\right)^2 = \frac{MG}{R}$$
$$\frac{4\pi^2 R^2}{T^2} = \frac{MG}{R}$$
$$T^2 = \frac{4\pi^2}{GM} R^3$$

The above expression is known as Kepler's third law for circular orbits. Solve for the mass of Eros.

$$M = \frac{4\pi^2 R^3}{T^2}$$
$$= 6.65 \times 10^{12} \text{ kg}$$

P43:
 Solution:

(a)

$$T = \frac{6.88 \text{ s}}{10 \text{ revolutions}} = 0.688 \text{ s}$$

$$|\vec{v}| = \frac{2\pi R}{T}$$
$$= 13.7 \text{ m/s}$$

(b) Yes, the momentum vector changes because its direction changes. It is not moving in a straight line.

(c) The force by the spring on the ball changes its direction because this force is perpendicular to the ball's path.

(d) $\left|\vec{F}_{\text{spring}}\right| = k_{\text{s}} = 300$ N

(e)

$$\left|\vec{F}_{\text{net}}\right| = \frac{m|\vec{v}|^2}{R}$$
$$= 300 \text{ N}$$

Solving for mass gives $m = 32.8$ kg.

P49:
 Solution:

The rider's speed is

$$|\vec{v}| = \frac{2\pi R}{T}$$
$$= 5.98 \text{ m/s}$$

The rider's speed is constant; therefore, the net force on the rider is always directed toward the center of the Ferris wheel and has a magnitude

$$\left|\vec{F}_{\text{net},\perp}\right| = |\vec{p}|\frac{|\vec{v}|}{R}$$
$$= \frac{m|\vec{v}|^2}{R} \text{ since the rider's speed is much less than c.}$$
$$= 200 \text{ N}$$

At each point in the motion, sketch a free-body diagram. There are two forces acting on the rider: the force by the seat and the gravitational force by Earth. The gravitational force by Earth always acts downward. The force by the seat on the varies so that when added to the gravitational force, the net force points toward the center of the Ferris wheel and has a constant magnitude.

The gravitational force by Earth on the rider is $\langle 0, -mg, 0 \rangle = \langle 0, -549, 0 \rangle$ N.

In all answers below, define the +y direction to be upward and the +x direction to the right.

(a) $\frac{d\vec{p}}{dt} = \langle 0, 200, 0 \rangle$ kg$\cdot$m/s^2

(b) $\langle 0, -549, 0 \rangle$ N

(c) The net force on the rider is

$$\vec{F}_{net} = \vec{F}_{grav} + \vec{F}_{seat}$$

Solve for the force by the seat on the rider.

$$
\begin{aligned}
\vec{F}_{seat} &= \vec{F}_{net} - \vec{F}_{grav} \\
&= \langle 0, 200, 0 \rangle \text{ N} - \langle 0, -549, 0 \rangle \text{ N} \\
&= \langle 0, 749, 0 \rangle \text{ N}
\end{aligned}
$$

(d) $\frac{d\vec{p}}{dt} = \langle 0, -200, 0 \rangle$ kg$\cdot$m/s^2

(e) $\langle 0, -549, 0 \rangle$ N

(f)

$$
\begin{aligned}
\vec{F}_{net} &= \vec{F}_{grav} + \vec{F}_{seat} \\
\vec{F}_{seat} &= \vec{F}_{net} - \vec{F}_{grav} \\
&= \langle 0, -200, 0 \rangle \text{ N} - \langle 0, -549, 0 \rangle \text{ N} \\
&= \langle 0, 349, 0 \rangle \text{ N}
\end{aligned}
$$

(g) The rider feels heavier at the bottom of the ride because $\left| F_{seat,y} \right| > \left| \vec{F}_{grav} \right|$.

(h) The rider feels lighter at the top of the ride because $\left| F_{seat,y} \right| < \left| \vec{F}_{grav} \right|$.

P51:
Solution:

A circular pendulum is analyzed as an example in CH 05 of the textbook. The radius of the pendulum is $R = L \sin\theta = 0.516$ m. Application of the Momentum Principle in the vertical direction shows that

$$F_T \cos\theta = mg$$

Application of the Momentum Principle in the radial direction shows that

$$F_T \sin\theta = \frac{m|\vec{v}|^2}{R}$$

Solving for F_T in the first equation and substituting into the second equation allows one to solve for the speed of the pendulum.

$$
\begin{aligned}
mg\tan\theta &= \frac{m|\vec{v}|^2}{R} \\
g\tan\theta &= \frac{|\vec{v}|^2}{R} \\
|\vec{v}| &= \sqrt{Rg\tan\theta} \\
&= \sqrt{(0.516 \text{ m})(9.8 \text{ N/kg})\tan(28°)} \\
&= 2.69 \text{ m/s}
\end{aligned}
$$

The period is found using:

$$
\begin{aligned}
|\vec{v}| &= \frac{2\pi R}{T} \\
T &= 1.21 \text{ s}
\end{aligned}
$$

P55:

Solution:

(a) The period of the star can be found from the figure. From 1995 to 2004 the star travels approximately 90 degrees, or 1/4 of a revolution around the circle. Therefore, the period of the star is $T = 4(9 \text{ y}) = 36$ y. Convert this to seconds, $T = 1.136 \times 10^9$ s.

The speed of the star is

$$
\begin{aligned}
|\vec{v}| &= \frac{2\pi r}{T} \\
&= \frac{2\pi(2.9 \times 10^{14})}{1.136 \times 10^9 \text{ s}} \\
&= 1.60 \times 10^6 \text{ m/s}
\end{aligned}
$$

In terms of the speed of light, the star's speed is $(1.60 \times 10^6 \text{ m/s})/(3 \times 10^8 \text{ m/s}) = 0.005c$

(b) The star's speed is less than one-hundredth the speed of light. As a result, it is considered non-relativistic, and $p \approx mv$.

(c) Use Kepler's third law.

$$
\begin{aligned}
T^2 &= \frac{4\pi^2}{GM}r^3 \\
M &= 1.11 \times 10^{37} \text{ kg}
\end{aligned}
$$

Divide this by the mass of Sun to get the mass in units of solar masses. $(1.11 \times 10^{37}$ kg$)/(2 \times 10^{30}$ kg$) = 5.6$ million solar masses.

CP57:
Solution:

(a) See the VPython program listing below. The yellow arrow represents the net force on the spacecraft.

(b) See the VPython program listing below.

(c) It is easiest to use the dot product of $\vec{F}_{net} \bullet \hat{p}$. VPython has a function dot(A,B) which calculates the dot product of two vectors A and B. The parallel component of the net force is $\left|\vec{F}_{net}\right| \cos\theta$. Thus, it can be calculated using

$$\vec{F}_{net,\parallel} = \left|\vec{F}_{net}\right| \cos\theta\hat{p}$$
$$= \frac{\vec{F}_{net} \cdot \vec{p}}{|\vec{p}|}\hat{p}$$

The net force can be written as $\vec{F}_{net} = \vec{F}_{\perp} + \vec{F}_{\parallel}$. Use this to solve for the perpendicular component of the net force.

$$\vec{F}_{\parallel} = \vec{F}_{net} - \vec{F}_{\perp}$$

The green arrow represents the force component perpendicular to the momentum. The red arrow represents the force component parallel to the momentum.

(d) The parallel component of force points in the same direction of the momentum when the spacecraft is on the inbound half of its orbit. The effect is to increase the spacecraft's speed and therefore also the magnitude of its momentum.

(e) The parallel component of force points opposite the momentum when the spacecraft is on the outbound half of its orbit. The effect is to decrease the spacecraft's speed and therefore also the magnitude of its momentum.

(f) The parallel component of force is zero at the apsides (rhymes with "rhapsodies") of the orbit, that is as the points closest to, and farthest from, Earth's center.

Here is the program listing.

```
from __future__ import division, print_function
from visual import *

RE = 6.4e6 #radius of Earth

spacecraft = sphere(pos=(-10*RE,0,0), color=color.cyan, radius=0.25*RE)
Earth = sphere(color=color.blue, radius=RE)

m=1.5e4 #mass of spacecraft
ME = 6e24
G = 6.67e-11

v=1.2*sqrt(G*ME/mag(spacecraft.pos))*vector(0,1,0) #initial v of spacecraft
p=m*v #initial p of spacecraft
```

```
t=0
dt=0.1*3600 #time step

rmag=mag(spacecraft.pos); #distance of spacecraft from Earth

trail=curve(color=spacecraft.color)
Fperparrow = arrow(pos=spacecraft.pos, axis=vector(0,0,0), color=color.green)
Fpararrow = arrow(pos=spacecraft.pos, axis=vector(0,0,0), color=color.red)
Fnetarrow = arrow(pos=spacecraft.pos, axis=vector(0,0,0), color=color.yellow)

scale = 10*RE/1000
while rmag>RE: #stop if rmag < RE
    rate(100)
    r=spacecraft.pos
    rmag=mag(r)
    runit=r/rmag

    Fgrav=-G*m*ME/rmag**2*runit
    Fnet=Fgrav

    p = p + Fnet*dt
    v = p/m
    spacecraft.pos = spacecraft.pos + v*dt

    #calculate the parallel component of the net force
    pmag = mag(p)
    phat = p/pmag
    Fparallel = dot(Fnet,p)/pmag*phat

    #calculate the perpendicular component of the net force
    Fperp = Fnet - Fparallel

    #update arrows for the forces
    Fperparrow.pos = spacecraft.pos
    Fperparrow.axis = Fperp*scale
    Fpararrow.pos = spacecraft.pos
    Fpararrow.axis = Fparallel*scale
    Fnetarrow.pos = spacecraft.pos
    Fnetarrow.axis = Fnet * scale

    trail.append(pos=spacecraft.pos)

    t = t + dt
```

Here is a screenshot of the program.

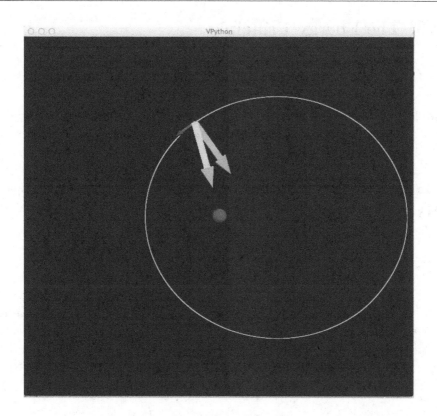

6   Chapter 6: The Energy Principle

Q03:

Solution:

You do the same work on each block. According to the Energy Principle, each block will have the same ΔK. Since they both start from rest, then K_f is the same for each block.

$K_f = \frac{p_f^2}{2m}$ therefore the larger mass block will have a larger final momentum.

If you pull the blocks for the same amount of time, then according to the Momentum Principle, $\Delta \vec{p} = \vec{F}_{net} \Delta t$ will be the same for the two blocks. Since they both start from rest, $\vec{p}_f$ will be the same. Since $\vec{v}_f = \frac{\vec{p}_f}{m}$, then $\left|\vec{p}_f\right|$ is greater for the smaller block and the smaller block will have a greater final kinetic energy.

Q07:

Solution:

To be associated with a potential energy function, the work done by a force around a path from A to B must be path independent. (There are other criteria too, but if a force fails this test, you know that it has no potential energy function.) Chose point A to be at the top of the circle. Choose point B at the bottom of the circle. If you select a path clockwise from A to B, then

$$\text{clockwise;} \quad W = \int_i^f \vec{F} \cdot d\vec{r}$$
$$= +\pi R F$$

where R is the radius and $\vec{F} \cdot d\vec{r} = +F\,dr$ since the force is in the same direction as a small displacement $d\vec{r}$. If you select a path counterclockwise from A to B, then

$$\text{clockwise;} \quad W = \int_i^f \vec{F} \cdot d\vec{r}$$
$$= -\pi R F$$

where R is the radius and $\vec{F} \cdot d\vec{r} = -F\,dr$ since the force is in the opposite direction as a small displacement $d\vec{r}$.

Because the integral depends on the path, then the force does not have an associated potential energy function.

P11:

Solution:

Because the energy of the electron is so large compared to its rest energy, the electron's speed will be very close to the speed of light. To achieve at least 7 significant figures in your answer, you need to use $m_{electron} = 9.10938215 \times 10^{-31}$ kg, $1.672621638 \times 10^{-27}$ kg and $c = 2.99792458 \times 10^8$ m/s.

$$E_{rest,proton} = mc^2$$
$$= (1.672621638 \times 10^{-27} \text{ kg})(2.99792458 \times 10^8 \text{m/s})^2$$
$$= 1.50327736 \times 10^{-10} \text{ J}$$

$E_{\text{electron}} = 1.50327736 \times 10^{-10}$ J and $m = 9.10938215 \times 10^{-31}$ kg

$$
\begin{aligned}
E_{\text{electron}} &= \gamma m c^2 \\
\gamma &= \frac{1.50327736 \times 10^{-10} \text{ J}}{(9.10938215 \times 10^{-31} \text{ kg})(2.99792458 \times 10^8 \text{ m/s})^2} \\
&= 1836.1527 \\
\gamma &= \frac{1}{\sqrt{1 - \frac{|\vec{v}|^2}{c^2}}}
\end{aligned}
$$

Solve for $|\vec{v}|$. Because of the very high value of γ, the speed of the electron is very close to c. In fact, the precision of your calculator may be so small, that it calculates $|\vec{v}|$ to be exactly equal to c, which is not correct. It's possible to use a computer program such as Excel which has greater precision than your calculator. The result is $|\vec{v}|/c = 0.99999985$, though the 5 is not a significant figure.

Note that most of the electron's energy is kinetic energy.

$$
\begin{aligned}
K &= (\gamma - 1) m c^2 \\
&= (1836.1527 - 1)(9.10938215 \times 10^{-31} \text{ kg})(2.99792458 \times 10^8 \text{ m/s})^2 \\
&= 1.502 \times 10^{-10} \text{ J}
\end{aligned}
$$

P15:
Solution:

$$
\begin{aligned}
K &= \frac{1}{2} m |\vec{v}|^2 \\
&= \frac{1}{2}(0.144 \text{ kg})(22^2 + 23^2 + 11^2 \text{ m}^2/\text{s}^2) \\
&= 81.6 \text{ J}
\end{aligned}
$$

P21:
Solution:

$$
\begin{aligned}
W &= \vec{F} \bullet \Delta \vec{r} \\
W &\approx \left\langle 0, -\left(9.8 \,\frac{\text{N}}{\text{kg}}\right)(2 \text{ kg}), 0 \right\rangle \text{ N} \bullet \langle 25, -30, 0 \rangle \text{ m} \\
W &\approx 585 \text{ J}
\end{aligned}
$$

P25:
Solution:

$$\begin{aligned}
\vec{F} \cdot \Delta\vec{r} &= (< 23, -12, 32 > \text{ N}) \cdot (< 0.12, 0.31, -0.24 > \text{ m}) \\
&= (23 \text{ N})(0.12 \text{ m}) + (-12 \text{ N})(0.31 \text{ m}) + (32 \text{ N})(-0.24 \text{ m}) \\
&= -8.6 \text{ J}
\end{aligned}$$

P31:
 Solution:

$$\begin{aligned}
E_{\text{rest}} &= mc^2 \\
&= 4.59 \times 10^{-10} \text{ J}
\end{aligned}$$

$E = \gamma mc^2$ where $\gamma = \dfrac{1}{\sqrt{1 - \frac{|\vec{v}|^2}{c^2}}} = 5.61$.

$$\begin{aligned}
E &= (5.61) E_{\text{rest}} \\
&= 2.58 \times 10^{-9} \text{ J}
\end{aligned}$$

$$\begin{aligned}
K &= E - E_{\text{rest}} \\
&= 2.12 \times 10^{-9} \text{ J}
\end{aligned}$$

If $W = 4.7 \times 10^{-9}$ J, then $\Delta E = W = 4.7 \times 10^{-9}$ J.

$$\begin{aligned}
E_{\text{f}} &= E_{\text{i}} + \Delta E \\
&= 2.58 \times 10^{-9} \text{ J} + 4.7 \times 10^{-9} \text{ J} \\
&= 7.28 \times 10^{-9} \text{ J}
\end{aligned}$$

E_{rest} stays the same, 4.59×10^{-10} J.

$$\begin{aligned}
K &= E - E_{\text{rest}} \\
&= 7.28 \times 10^{-9} \text{ J} - 4.59 \times 10^{-10} \text{ J} \\
&= 6.82 \times 10^{-9} \text{ J}
\end{aligned}$$

Since E_{rest} stays the same, you can also calculate K using $K_{\text{f}} = K_{\text{i}} + W = 6.82 \times 10^{-9}$ J

P37:

    Solution:

$$
\begin{aligned}
W &= \vec{\mathbf{F}} \bullet \Delta \vec{\mathbf{r}} \\
&= \langle 250, 490, -160 \rangle \text{ N} \bullet \langle 3, -9, -5 \rangle \text{ m} \\
&= -2860 \text{ J}
\end{aligned}
$$

$$
\begin{aligned}
K_f &= K_i + W \\
&= \frac{1}{2} m v_i^2 + W \\
&= 8640 \text{ J} + (-2860 \text{ J}) \\
&= 5780 \text{ J}
\end{aligned}
$$

$$
\begin{aligned}
K_i &= \frac{1}{2} m v_f^2 \\
v_f &= 9.81 \text{ m/s}
\end{aligned}
$$

P41:

    Solution:

Steps 1, 2, 5, and 6 must be included, but not necessarily in that order.

$$
\begin{aligned}
W &= \vec{\mathbf{F}} \bullet \Delta \vec{\mathbf{r}} \\
&\approx \left\langle 1.6 \times 10^{-13}, 0, 0 \right\rangle \text{ N} \bullet \langle 2, 0, 0 \rangle \text{ m} \\
&\approx 3.2 \times 10^{-13} \text{ J} \\
E_f &= E_i + W \\
&\approx \frac{1}{\sqrt{1 - \frac{0.91 c^2}{c^2}}} (9 \times 10^{-31} \text{ kg})(3 \times 10^8 \, \frac{\text{m}}{\text{s}})^2 + 3.2 \times 10^{-13} \text{ J} \\
&\approx 5.15 \times 10^{-13} \text{ J}
\end{aligned}
$$

Solve for speed as a function of energy (final energy, that is).

$$E = \gamma m c^2$$

$$\gamma = \frac{E}{mc^2} = \frac{1}{\sqrt{1 - \frac{|\vec{v}|^2}{c^2}}}$$

$$\frac{|\vec{v}|}{c} = \sqrt{1 - \frac{mc^2}{E}}$$

$$\frac{|\vec{v}|}{c} \approx \sqrt{1 - \frac{(9 \times 10^{-31}\ \text{kg})(3 \times 10^8\ \frac{\text{m}}{\text{s}})^2}{5.15 \times 10^{-13}\ \text{J}}}$$

$$\frac{|\vec{v}|}{c} \approx 0.99$$

Significant figures are important in this problem.

P45:

Solution:

(a)

$$E_{i,\text{rest}} = mc^2$$
$$= (3.894028 \times 10^{-25}\ \text{kg})(2.99792 \times 10^8\ \text{m/s})^2$$
$$= 3.499767 \times 10^{-8}\ \text{J}$$

(b)

$$E_{rest,alpha} + E_{\text{rest, new nucleus}} = m_\alpha c^2 + m_{\text{nucleus}} c^2$$
$$= (6.640678 \times 10^{-27}\ \text{kg})(2.99792 \times 10^8\ \text{m/s})^2 + (3.827555 \times 10^{-25}\ \text{kg})(2.99792 \times 10^8\ \text{m/s})$$
$$= 5.968326 \times 10^{-10}\ \text{J} + 3.440024 \times 10^{-8}\ \text{J}$$
$$= 3.499708 \times 10^{-8}\ \text{J}$$

(c) The rest energy decreased.

(d)

$$K = E_{\text{rest,i}} - E_{\text{rest,f}}$$
$$= 5.95 \times 10^{-12}\ \text{J}$$
$$= 3.72\ \text{MeV}$$

P53:

Solution:

The system is the ball and Earth.

$$U_i + K_i = U_f + K_f$$

$$mgy_i + \frac{1}{2}mv_i^{2\,0} = mgy_f^{\,0} + \frac{1}{2}mv_f^2$$

$$\cancel{m}gy_i = \frac{1}{2}\cancel{m}v_f^2$$

$$v_f = \sqrt{2gy_i}$$

$$= 6.3\ \text{m/s}$$

P59:

  Solution:

(a)

$$r_i = 2 \times 10^6\ \text{m}$$
$$r_f = \infty\ \text{i.e. very far away}$$
$$M = 1.2 \times 10^{23}\ \text{kg}$$
$$v_f = 900\ \text{m/s}$$
$$v_i = ?$$
$$m = ?$$

Define the planet and object as the system. Assume it is closed system, so energy is conserved. Since rest energy does not change and since the planet's kinetic energy is negligible (due to its extremely small speed), the energy of the system is

$$E_{sys} = K_{object} + U_{grav,\ sys}$$

Choose the initial point to be when the object is launched from the surface. Choose the final point to be very far away from the planet (i.e. $r = \infty$). Then

$$E_i = E_f$$
$$U_{grav,\ i} + K_i = U_{grav,\ f} + K_f$$
$$\frac{-GmM}{r_i} + \frac{1}{2}mv_i^2 = \frac{-GmM}{r_f} + \frac{1}{2}mv_f^2$$

When the object is far away, $r \approx \infty$ and $1/r \approx 0$. Thus

$$\frac{-GmM}{r_i} + \frac{1}{2}mv_i^2 = \frac{1}{2}mv_f^2$$

$$\frac{1}{2}mv_i^2 = \frac{1}{2}mv_f^2 + \frac{GmM}{r_i}$$

$$\frac{1}{2}\cancel{m}v_i^2 = \frac{1}{2}\cancel{m}v_f^2 + \frac{G\cancel{m}M}{r_i}$$

$$v_i^2 = v_f^2 + \frac{2GM}{r_i}$$

$$v_i = \left(v_f^2 + \frac{2GM}{r_i}\right)^{1/2}$$

$$= \left((900 \text{ m/s})^2 + \frac{2\left(6.7 \times 10^{-11} \frac{\text{Nm}^2}{\text{kg}^2}\right)(1.2 \times 10^{23} \text{ kg})}{2 \times 10^6 \text{ m}}\right)^{1/2}$$

$$= 2970 \text{ m/s}$$

(b) To calculate the initial speed to just "escape," set the final speed in the equation for part (a) to zero. Then the required initial speed is known as the escape speed.

$$= \left((\frac{2\left(6.7 \times 10^{-11} \frac{\text{Nm}^2}{\text{kg}^2}\right)(1.2 \times 10^{23} \text{ kg})}{2 \times 10^6 \text{ m}}\right)^{1/2}$$

$$= 2835 \text{ m/s}$$

P65:

Solution:

$$W = \left|\vec{F}\right|\left|\vec{\Delta r}\right|\cos\theta = (1.6 \times 10^{-13} \text{ N})(2 \text{ m}) = 3.2 \times 10^{-13} \text{ J}$$

$$W = \Delta E$$

$$= E_f - E_i$$

$$E_f = W + E_i$$

$$\gamma_f mc^2 = W + \gamma_i mc^2$$

$$= 3.2 \times 10^{-13} \text{ J} + \frac{1}{\sqrt{1 - (0.95)^2}}(9.11 \times 10^{-31} \text{ kg})(3 \times 10^8 \text{ m/s})^2$$

$$= 3.2 \times 10^{-13} \text{ J} + 2.63 \times 10^{-13} \text{ J}$$

$$= 5.83 \times 10^{-12} \text{ J}$$

$$\gamma_f mc^2 = 5.83 \times 10^{-13} \text{ J}$$

$$\gamma_f = \frac{5.83 \times 10^{-13} \text{ J}}{(9.11 \times 10^{-31} \text{ kg})(3 \times 10^8 \text{ m/s})^2}$$

$$\gamma_f = 7.11$$

$$\gamma_f = \frac{1}{\sqrt{1 - \frac{v^2}{c^2}}}$$

Solve for v which gives $v_f = 0.99c$.

P69:
   Solution:

(a) Define $y = 0$ to be at the lowest point of the pendulum's motion ($\theta = 0$). Then the height of the pendulum is $y = L - L\cos\theta = L(1 - \cos\theta)$. Because it is near Earth, we can use

$$
\begin{aligned}
U &= mgy \\
U &= mgL(1 - \cos\theta)
\end{aligned}
$$

(b) To sketch the graph, assume that $m = 1$ kg and $L = 1$ m. Then, U varies between the values of 0 and $2g$. An example graph is shown in the figure below.

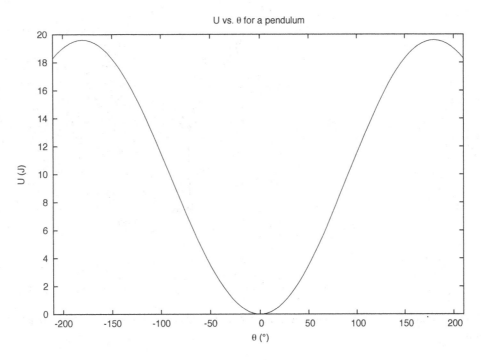

(c) $s = L\theta$, so $\theta = \frac{s}{L}$. Write U in terms of s.

$$
U = mgL(1 - \cos\frac{s}{L})
$$

The tangential component of the net force on the mass is

$$
\begin{aligned}
F_{net,tan} &= -\frac{\partial U}{\partial s} \\
&= -\frac{\partial}{\partial s}\left(mgL\left(1 - \cos\frac{s}{L}\right)\right) \\
&= mgL\left(-\sin\frac{s}{L}\right)\left(\frac{1}{L}\right) \\
&= -mg\sin\frac{s}{L} \\
&= -mg\sin\theta
\end{aligned}
$$

The net force can also be found by sketching a free-body diagram for the pendulum and summing the forces to calculate the net force on the pendulum. A free-body diagram is shown in the figure below.

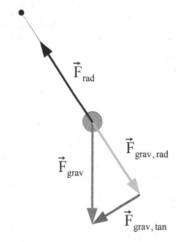

The gravitational force can be written in terms of its radial component and its tangential component. The tangential component of $\vec{F}_{grav}$ is

$$
\begin{aligned}
F_{grav,tan} &= -F_{grav}\sin\theta \\
&= -mg\sin\theta
\end{aligned}
$$

The negative sign gives the direction, showing that the force is to the "left" and angent to the path for positive angles and to the "right" and tangent to the path for negative angles.

(d) To barely make it to the top, $v_f = 0$ at the top. Apply the Energy Principle with the initial state at $\theta = 0$ and the final state at $\theta = 180°$.

$$
\begin{aligned}
E_i &= E_f \\
K_i &= U_f \\
K_i &= mgL(1 - \cos 180°) \\
\frac{1}{2}mv_i^2 &= 2mgL \\
v_i &= \sqrt{4gL}
\end{aligned}
$$

Use $m = 1$ kg and $L = 1$ m. The initial energy needed to barely make it to the top is $E_i = 2mgL = 2g$, for $m = 1$ and $L = 1$. If $E > 2g$, then it easily goes all the way around. If $E < 2g$, then it never makes it to the top. An example graph is shown in the figure below.

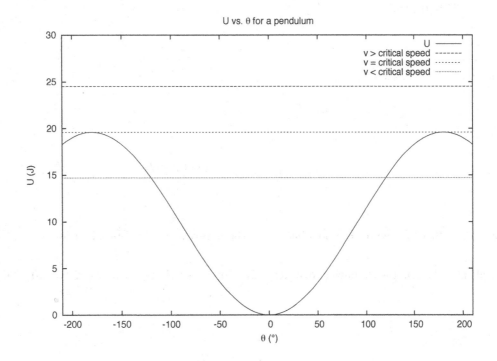

P75:
   Solution:

(a)

$$E_i = E_f$$
$$E_{rest,U} = 2E_{rest,Pd} + 2K_{Pd}$$
$$m_U c^2 = 2m_{Pd} c^2 + 2K_{Pd}$$
$$K_{Pd} = \frac{1}{2}(m_U c^2 - 2m_{Pd} c^2)$$
$$= \frac{1}{2}\left((235.996)(1.6603 \times 10^{-27}\text{ kg})(2.99792 \times 10^8\text{ m/s})^2 - 2(117.894)(1.6603 \times 10^{-27}\text{ kg})(2.99792 \times 10^8\text{ m/s})^2\right)$$
$$= \frac{1}{2}(3.52153 \times 10^{-8}\text{ J} - 3.51843 \times 10^{-8}\text{ J})$$
$$= \frac{1}{2}(3.103 \times 10^{-11}\text{ J})$$
$$= 1.55 \times 10^{-11}\text{ J}$$

Assuming that $|\vec{v}| << c$, then

$$K = \frac{1}{2}m|\vec{v}|^2$$
$$v_{Pd} = 1.26 \times 10^7\text{ m/s}$$

This is less than 10% of c, so it is reasonable to use the non-relativistic approximation.

(b) Immediately after fission, the Pd nuclei are at rest. Each one has a charge of $46(1.60218 \times 10^{-19}\text{ C}) = 7.37003 \times 10^{-18}\text{ C}$.

$$E_i = E_f$$
$$U_i = 2K_{f,Pd}$$
$$\frac{1}{4\pi\epsilon_0}\frac{q_1 q_2}{r_i} = 3.103 \times 10^{-11}\text{ J}$$
$$r_i = 1.57 \times 10^{-14}\text{ m}$$

(c)

$$R = (1.3 \times 10^{-15}\text{ m})(118)^{\frac{1}{3}}$$
$$= 6.38 \times 10^{-15}\text{ m}$$

The distance between the centers of the nuclei is 15.7×10^{-15} m. Thus the gap between the surfaces is $15.7 - 2(6.38) = 2.95 \times 10^{-15}$ m. A sketch is shown in the figure below. Note that 10^{-15} m = 1 fm.

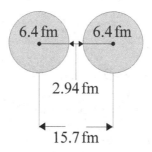

(d) Each U atom produces 2 Pd atoms with $2(1.55 \times 10^{-11}$ J of kinetic energy. The total kinetic energy produced by 1 mol of U atoms is

$$\left(\frac{6.023 \times 10^{23} \text{ U atoms}}{\text{mol}}\right)\left(\frac{2(1.55 \times 10^{-11} \text{ J})}{\text{U atom}}\right) \;=\; 1.9 \times 10^{13} \text{ J}$$

CP77:
Solution:

(a) Here is a sample program that plots K, U, and E.

```
from __future__ import division, print_function
from visual import *
from visual.graph import *

RE = 6.4e6 #radius of Earth

spacecraft = sphere(pos=(-10*RE, 0, 0), color=color.cyan, radius=0.25*RE, make_trail
    =true)
Earth = sphere (color=color.blue, radius=RE)

m=1.5e4 #mass of spacecraft
ME = 6e24
G = 6.67e-11

v=vector(0,3.27e3,0) #initial velocity of spacecraft
p=m*v #initial momentum of spacecraft

t=0
dt=60 #time step

rmag=mag(spacecraft.pos); #distance of spacecraft from Earth
```

```
Parrow=arrow(pos=spacecraft.pos, axis=(0,0,0), color=color.yellow)
Farrow=arrow(pos=spacecraft.pos, axis=(0,0,0), color=color.magenta)
Fscale=12*RE/1466
Pscale=12*RE/4.5e7

# initialize graph display
graph1 = gdisplay(x=430, y=0, width=430, height=450,
                  title='Energy vs. Time',
                  xtitle='t (s)',ytitle='E (J)')

# initialize the function to be plotted
Kplot = gcurve(color=color.cyan)
Uplot = gcurve(color=color.green)
Eplot = gcurve(color=color.yellow)


while rmag>RE: #stop if rmag < RE
    rate(1000)
    r=spacecraft.pos
    rmag=mag(r)
    runit=r/rmag

    Fgrav=-G*m*ME/rmag**2*runit
    Fnet=Fgrav

    p = p + Fnet*dt
    v = p/m
    spacecraft.pos = spacecraft.pos + v*dt

#    print(mag(Fnet), mag(p))
    Farrow.pos=spacecraft.pos
    Farrow.axis=Fscale*Fnet
    Parrow.pos=spacecraft.pos
    Parrow.axis=Pscale*p

    U=-G*m*ME/mag(r)
    K=0.5*m*mag(v)**2
    E=U+K

    # update graph
    Kplot.plot(pos=(t,K))
    Uplot.plot(pos=(t,U))
    Eplot.plot(pos=(t,E))

    t = t + dt
```

The output is shown below.

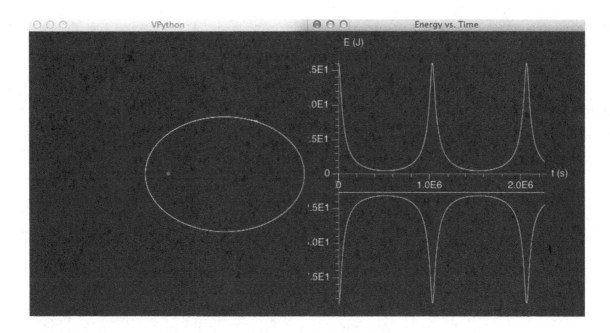

The total energy is constant and negative (as it should be for a bound system). The kinetic energy is always positive (as it should be), and the gravitational potential energy is always negative (as it should be). When the kinetic energy increases, the gravitational potential energy decreases. And when the kinetic energy decreases, the gravitational potential energy increases. This occurs because E is constant.

(b) In Problem P65 in Chapter 3, we found that an approximate maximum value of Δt was 1 hour (i.e. 3600 s). So begin by testing this to see if E remains constant on the graph. You'll notice some spikes when the spaceship passes closest to Earth (called the perigee), as shown below.

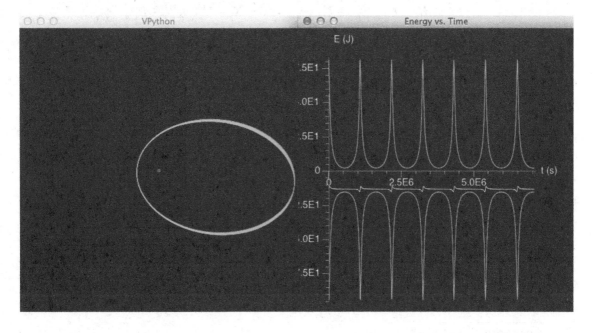

Trying 0.5 h still shows spikes in the total energy. Even 0.2 h shows spikes. However, 0.1 h shows an almost imperceptibly small bump in the total energy curve. It seems that $\Delta t = 360$ s is about the smallest possible time step that should be used for an accurate model.

(c) If you decrease the speed to 3×10^4 m/s, the total energy decreases. The closer that the orbit gets to being a circle (i.e. less eccentric), the bottom of the kinetic energy curve and the peak of the potential energy curve get broader and flatter. For a circular orbit, the graphs of K and U become flat.

CP83:
Solution:

For both the VPython program and the analytical calculation, I will use the following constants.

$$
\begin{aligned}
v_i &= 1 \times 10^4 \text{ m/s} \\
h &= 5 \times 10^4 \text{ m} \\
R_E &= 6.4 \times 10^6 \text{ m} \\
R_M &= 1.75 \times 10^6 \text{ m} \\
m &= 173 \text{ kg} \\
m_E &= 6 \times 10^{24} \text{ kg} \\
m_M &= 7 \times 10^{22} \text{ kg} \\
G &= 6.7 \times 10^{-11} \frac{\text{Nm}^2}{\text{kg}^2} \\
r &= 4 \times 10^8 \text{ m} \qquad \text{center-to-center distance between Earth and Moon}
\end{aligned}
$$

Define the system as Ranger, Earth, and Moon. Apply the Energy Principle. There are two pairs of interacting objects: Ranger-Earth and Ranger-Moon. Thus, there are two potential energy terms.

$$
\begin{aligned}
E_i &= E_f \\
K_i + U_{grav,R,E,i} + U_{grav,R,M,i} &= K_f + U_{grav,R,E,f} + U_{grav,R,M,f} \\
\frac{1}{2}mv_i^2 + \frac{-GmM_E}{r_{\text{E to R,i}}} + \frac{-GmM_M}{r_{\text{M to R,i}}} &= \frac{1}{2}mv_f^2 + \frac{-GmM_E}{r_{\text{E to R,f}}} + \frac{-GmM_M}{r_{\text{M to R,f}}} \\
\frac{1}{2}mv_i^2 + \frac{-GmM_E}{R_E + h} + \frac{-GmM_M}{r - R_E - h} &= \frac{1}{2}mv_f^2 + \frac{-GmM_E}{r - R_M} + \frac{-GmM_M}{R_M}
\end{aligned}
$$

The mass of Ranger (m) cancels out, and you can solve for v_f.

$$
\begin{aligned}
v_f^2 &= v_i^2 + 2GM_E \left(\frac{1}{r - R_M} - \frac{1}{R_E + h} \right) + 2GM_M \left(\frac{1}{R_M} - \frac{1}{r - R_E - h} \right) \\
&= (1.3 \times 10^4 \text{ m/s})^2 + 2 \left(6.7 \times 10^{-11} \frac{\text{Nm}^2}{\text{kg}^2} \right) (6 \times 10^{24} \text{ kg}) \left(\frac{1}{4 \times 10^8 \text{ m} - 1.75 \times 10^6 \text{ m}} - \frac{1}{6.4 \times 10^6 \text{ m} + 5 \times 10^4 \text{ m}} \right) \\
&\quad + 2 \left(6.7 \times 10^{-11} \frac{\text{Nm}^2}{\text{kg}^2} \right) (7 \times 10^{22} \text{ kg}) \left(\frac{1}{1.75 \times 10^6 \text{ m}} - \frac{1}{4 \times 10^8 \text{ m} - 6.4 \times 10^6 \text{ m} - 5 \times 10^4 \text{ m}} \right) \\
&= 5.223 \times 10^7 \text{ m}^2/\text{s}^2 \\
v_i &= \sqrt{5.223 \times 10^7 \text{ m}^2/\text{s}^2} \\
v_i &\approx 7230 \text{ m/s}
\end{aligned}
$$

Now print the speed of Ranger as calculated from a VPython program. Here is the program that I used. In a previous problem, we found that the simulation was most accurate for $\Delta t \leq 5$ s, so I used 5 s as the time step.

```
from __future__ import division, print_function
from visual import *
from visual.graph import *


RE = 6.4e6 #radius of Earth
RM = 1.75e6 #radius of Moon
h = 5e4 #initial altitude of Ranger

Earth = sphere(pos=(0,0,0), color=color.blue, radius=5*RE)
Moon = sphere(pos=(4e8,0,0), color=color.white, radius=0.5*Earth.radius)
ranger = sphere(pos=(RE+h, 0, 0), color=color.cyan, radius=0.25*Earth.radius,
    make_trail=True)

m=173 #mass of ranger
ME = 6e24 #mass of Earth
MM = 7e22 #mass of Moon
G = 6.67e-11

v=vector(1.3e4,0, 0) #initial velocity of ranger
p=m*v #initial momentum of ranger

t=0
dt=5

# initialize graph display
graph1 = gdisplay(x=430, y=0, width=600, height=450,
                  title='Energy vs. Time',
                  xtitle='t (s)',ytitle='E (J)')

# initialize graph display
graph1 = gdisplay(x=430, y=0, width=600, height=450,
                  title='Energy vs. Distance',
                  xtitle='r (m)',ytitle='E (J)')

# initialize the function to be plotted
Kplot = gcurve(color=color.cyan, dot=True)
Uplot = gcurve(color=color.green, dot=True)
Eplot = gcurve(color=color.yellow, dot=True)


while 1: #stop if rmag < RE or rrelmoonmag < RM
    rate(500)

    #calculate Fgrav on ranger by Earth
    r=ranger.pos
    rmag=mag(r)
    runit=r/rmag
    FgravE=-G*m*ME/rmag**2*runit

    #calculate Fgrav on ranger by Moon
```

```
rrelmoon=ranger.pos - Moon.pos
rrelmoonmag=mag(rrelmoon)
rrelmoonunit=rrelmoon/rrelmoonmag
FgravM=-G*m*MM/rrelmoonmag**2*rrelmoonunit

#calculate net force
Fnet=FgravE + FgravM

#update momentum and position
p = p + Fnet*dt
v = p/m
ranger.pos = ranger.pos + v*dt

#break loop if ranger crashed into Moon or Earth
if (mag(ranger.pos)<RE or mag(ranger.pos - Moon.pos)<RM):
    vf=sqrt(2*K/m)
    print("Speed before impact =",vf," m/s")
    print("Speed after impact =",mag(v)," m/s")
    break

#update energy
U=-G*m*ME/mag(ranger.pos) + -G*m*MM/mag(ranger.pos-Moon.pos)
K=0.5*m*mag(v)**2
E=U+K

# update graph
Kplot.plot(pos=(mag(ranger.pos),K))
Uplot.plot(pos=(mag(ranger.pos),U))
Eplot.plot(pos=(mag(ranger.pos),E))


t = t + dt
```

Let's look at the program. In the `while` loop, the position and velocity of Ranger is updated. When Ranger moves past the surface of Moon, we break the loop and print the speed. Because the velocity was already updated, the speed that is printed is just after the impact (assuming that it did not slow down due to impact). We can print the speed of Ranger at the previous step (just before impact) by calculating it from the kinetic energy (which is updated after the break statement).

Using the program above, we calculate the speed after impact to be 7180 m/s. This is similar to the speed of 7230 m/s that we calculated "by hand." In fact, they are only 0.7% different. It demonstrates that a computer program can be an accurate method to solve a problem.

7   Chapter 7: Internal Energy

Q07:
Solution:

(a) If the system is just the box, then I and Earth exert significant forces on the system.

(b) If the system is the box and me, then only Earth exerts significant force on the system.

(c) If the system is the box, Earth, and me, then nothing in the surroundings exerts a significant force on the system.

Q13:
Solution:

P19:
Solution:

(a) system = spring + block

$$\Delta E_{sys} = \cancelto{0}{\Delta K} + \Delta U_s = W_{hand} + W_{Earth}$$

$$\frac{1}{2}k_s s^2 = W_{hand} - Mgs, (s < 0)$$

$$W_{hand} = \frac{1}{2}k_s s^2 + Mgs$$

$$\text{But } Mg = k_s s$$

$$|s| = \frac{Mg}{k_s}$$

$$\therefore W_{hand} = \frac{1}{2}k_s \left(\frac{Mg}{k_s}\right)^2 + \frac{(Mg)^2}{k_s} = -\frac{1}{2}\frac{(Mg)^2}{k_s}$$

(b) system = spring + block + Earth

$$\cancelto{0}{\Delta K} + \Delta U_g + \Delta U_s = 0$$

$$Mgs + \frac{1}{2}k_s s^2 = 0, (s < 0)$$

$$\therefore s = -\frac{2Mg}{k_s}, (s < 0)$$

(c) system = spring + block + Earth

$$\Delta K + \Delta U_g + \Delta U_s = 0$$

$$\frac{1}{2}M\left|\vec{v}_f\right|^2 + Mg(2L) - Mg(L+s) - \frac{1}{2}k_s s^2 = 0$$

$$\left|\vec{v}_f\right| = \sqrt{\frac{k_s s^2}{M} - 2g(L-s)}$$

P23:
Solution:

Take the system to be spring, block, and Earth so that there are no significant external forces. Apply the momentum principle and solve for the spring's final height. This problem is very simple is you choose the proper system and if you reason in the following way.

The block will begin to fall, but it won't begin compressing the spring right away. The system will only lose gravitational potential energy, and therefore gain kinetic energy, as the block drops the first $d = 0.65$ m. Only then will it begin compressing the spring. At that moment, the system will begin losing the previously gained kinetic energy. The lost kinetic energy will show up at a gain in springy potential energy. As this happens, the block continues to drop though, and the system continues to lose gravitational potential energy. But there is a more efficient way to think about this problem, and that is to eliminate kinetic energy entirely by taking the initial state to be with the block at rest just after being released and the final state to be with the block at rest having fallen onto the spring, which is now compressed. The system has zero kinetic energy in both of these states, so the springy potential energy comes from the lost gravitational potential energy. The block drops through a total height equal to it's original height above the spring, d, plus the amount by which the spring is compressed, s. You can then solve for the compression, and then therefore the final height of the spring. You'll end up with a quadratic expression that must be solved.

$$E_{sys,i} = K_i + U_{s,i} + U_{g,i}$$
$$E_{sys,f} = K_f + U_{s,f} + U_{g,f}$$
$$\Delta E_{sys} = E_{sys,f} - E_{sys,i}$$
$$= \frac{1}{2}k_s s^2 - Mg(d+s) = 0$$
$$= s^2 - \frac{2Mg}{k_s}s - \frac{2Mgd}{k_s} = 0$$
$$s = \frac{\frac{2Mg}{k_s} \pm \sqrt{\left(\frac{2Mg}{k_s}\right)^2 + \frac{8Mg}{k_s}}}{2}$$
$$= \frac{Mg}{k_s} \pm \sqrt{\left(\frac{Mg}{k_s}\right)^2 + \frac{2Mgd}{k_s}}$$
$$\frac{Mg}{k_s} = \frac{(0.4 \text{ kg})(9.8 \text{ m/s}^2)}{(1000 \text{ N/m})} = 0.00392 \text{ m}$$
$$s = 0.00392 \text{ m} \pm \sqrt{(0.00392 \text{ m})^2 + 2(0.00392 \text{ m})(0.65 \text{ m})}$$
$$\therefore s = 0.0754 \text{ m}$$

Note that we discarded the negative root. So, in the final state the block will be 0.15 m $- 0.0754$ m $= 0.0746$ m above the floor.

P27:
Solution:

System is the water + pan. Assume it is insulated so $Q = 0$.

$$\Delta E_{sys} = 0$$
$$\Delta E_{therm,water} + \Delta E_{therm,Al} = 0$$
$$m_{water}\, c_{water}\, \Delta T_{water} + m_{Al}\, c_{Al}\, \Delta T_{Al} = 0$$
$$(400\text{ g})(4.2\ \frac{J}{K \cdot g})(T_f - 100^\circ C) + (600\text{ g})(0.9\ \frac{J}{K \cdot g})(T_f - 20^\circ C) = 0$$

Solve for T_f. You will find that $T_f = 80.5^\circ C$.

P33:
Solution:

The system is the water.

$$\Delta U_g + \Delta E_{therm} = 0$$
$$mg\Delta h + \Delta E_{therm} = 0$$
$$\Delta E_{therm} = mg\Delta h \approx (1\text{ kg})(9.8\ \frac{N}{kg})(50\text{ m})$$
$$\Delta E_{therm} = 490\text{ J}$$
$$P = \frac{\Delta E_{therm}/(1\text{ kg})}{1\text{ s}} M$$
$$M = \frac{1 \times 10^6\text{ J/s}}{490\text{ J/s}} = 2040\text{ kg}$$

P41:
Solution:

(a) $\vec{F}_{net} = 0$ at terminal speed, thus $\left|\vec{F}_{air}\right| = \left|\vec{F}_{grav}\right| = mg$.

So, $\left|\vec{F}_{air}\right| = 353\text{ N}$.

(b) $\left|\vec{F}_{air}\right| = \left|\vec{F}_{grav}\right| = 696\text{ N}$.

CP47:
Solution:

(a) Set $C = 0$ and run the simulation. The baseball lands at a distance of 204 m from its initial position at the clock reading $t = 6.45$ s. Now, solve this analytically. Due to symmetry, $v_{fy} = -v_{iy}$. Apply the Momentum Principle in the y-direction to calculate the time interval that the ball is in the air.

$$
\begin{aligned}
F_{net,y} &= \frac{\Delta p_y}{\Delta t} \\
-mg &= m\frac{v_{f,y} - v_{i,y}}{\Delta t} \\
-mg &= -m\frac{2v_{i,y}}{\Delta t} \\
g &= \frac{2v_{i,y}}{\Delta t} \\
\Delta t &= \frac{2v_{i,y}}{g} \\
&= \frac{2(44.7 \text{ m/s})\cos(45°)}{9.8 \text{ N/kg}} \\
&= 6.45 \text{ s}
\end{aligned}
$$

Note that the time matches the time determined in the simulation. Now, use this time to calculate the horizontal displacement of the baseball. The x-component of the net force is zero, so the x-velocity of the baseball is constant. Apply the definition of average velocity and solve for Δx.

$$
\begin{aligned}
v_{avg,x} &= \frac{\Delta x}{\Delta t} \\
\Delta x = v_{avg,x}\,\Delta t \\
&= (44.7 \text{ m/s})\sin(45°)(6.45 \text{ s}) \\
&= 204 \text{ m}
\end{aligned}
$$

This agrees with the results of the simulation.

(b) Run the program to see graphs of K, U, and $E = K + U$.

(c) Change C to its default nonzero value. The ball travels only 113 m downrange in 5.38 s.

(d) Energy is not conserved in the presence of air resistance.

(e) Substitute $\rho = 0.83(1.3 \text{ kg/m}^3) = 1.08 \text{ kg/m}^3$ into the simulation. The range in this case is 122 m. At sea level, it was 113 m. As a result, the ball "carries" 9 m, or 8%, further due to the lower density of air.

Here is the program.

```
from __future__ import division, print_function
from visual import *
from visual.graph import *

ball=sphere(radius=5, pos=vector(0,0,0), color=color.white)
grass=box(pos=vector(0,-ball.radius-0.5,0), length=250, width=250,
          height=1, color=color.green)

m=0.155 #mass in kg
s=100*0.44704 #initial speed in m/s
theta=45*pi/180. #initial angle in rad
v=s*vector(cos(theta),sin(theta),0) #velocity vector
p=m*v #momentum
```

```
g=9.8
#C=0 # use to ignore air resistance
C=0.35
R=0.07/2
A=pi*R**2
rho=1.3
rho=0.83*rho # use for Denver

dt=0.01
t=0.

graph=gdisplay(x=430,y=0,width=400, height=400,
            title='E vs t for object oscillating on a spring',
            xtitle='t (s)',
            ytitle='E (J)',
            background=color.black)

function=gcurve(gdisplay=graph, color=color.magenta)
function2=gcurve(gdisplay=graph, color=color.cyan)
function3=gcurve(gdisplay=graph, color=color.white)

trail=curve(color=ball.color)

scene.mouse.getclick()

while ball.pos.y>-0.01:
    rate(100)

    vmag=mag(v)
    vhat=v/vmag
    Fair = -1/2*C*rho*A*vmag**2*vhat
    Fgrav=vector(0,-m*g,0)

    Fnet=Fair+Fgrav
    p=p+Fnet*dt
    v=p/m
    ball.pos=ball.pos+v*dt
    t=t+dt

    U=m*g*ball.y
    K=1/2*m*vmag**2
    E=K+U

    function.plot(pos=(t,U))
    function2.plot(pos=(t,K))
    function3.plot(pos=(t,E))
    trail.append(pos=ball.pos)

print ("final position of the ball: ",ball.pos, "m at t=", t, "s")
print ("distance: ",mag(ball.pos), "m at t=", t, "s")
```

8   Chapter 8: Energy Quantization

Q01:

    Solution:

    (a) Arrow 1

    (b) Arrow 4

    (c) Arrow 2

    (d) Arrow 3

    (e) (2) through (5) are all correct.

P09:

    Solution:

    The atom's ground state energy is $-13.6\,\mathrm{eV}$ indicating a bound system. To make the total energy zero, we must add $13.6\,\mathrm{eV}$ and this ionizes the atom.

P15:

    Solution:

    (a) Two possible energy level schemes are shown in the figure below.

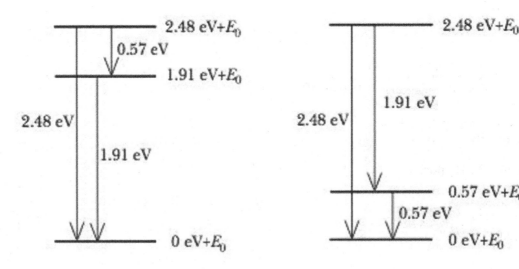

    (b) To distinguish between these two schemes, consider the absorption spectrum of a very cold quantum object, i.e. an object in the ground state. Bombard the object with photons with a variety of energies, but including an energy of $0.57\,\mathrm{eV}$. If we observe an $0.57\,\mathrm{eV}$ absorption line, the scheme on the right must be the correct one. Otherwise, the one on the left is the correct one.

P19:
 Solution:

First, sketch a diagram of the energy levels and the transitions that emit a photon. (See the figure below.)

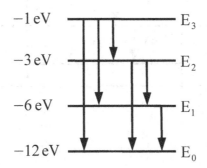

The emitted photons have the following energies:

$$
\begin{aligned}
\left|\Delta E_{0,3}\right| &= \left|E_0 - E_3\right| = \left|-12\,\text{eV} + 1\,\text{eV}\right| = 11\,\text{eV} \\[4pt]
\left|\Delta E_{1,3}\right| &= \left|E_1 - E_3\right| = \left|-6\,\text{eV} + 1\,\text{eV}\right| = 5\,\text{eV} \\[4pt]
\left|\Delta E_{2,3}\right| &= \left|E_2 - E_3\right| = \left|-3\,\text{eV} + 1\,\text{eV}\right| = 2\,\text{eV} \\[4pt]
\left|\Delta E_{0,2}\right| &= \left|E_0 - E_2\right| = \left|-12\,\text{eV} + 3\,\text{eV}\right| = 9\,\text{eV} \\[4pt]
\left|\Delta E_{1,2}\right| &= \left|E_1 - E_2\right| = \left|-6\,\text{eV} + 3\,\text{eV}\right| = 3\,\text{eV} \\[4pt]
\left|\Delta E_{0,1}\right| &= \left|E_0 - E_1\right| = \left|-12\,\text{eV} + 6\,\text{eV}\right| = 6\,\text{eV}
\end{aligned}
$$

For absorption when all atoms are in the ground state, the only transitions possible are $\left|\Delta E_{1,0}\right| = 6\,\text{eV}$, $\left|\Delta E_{2,0}\right| = 9\,\text{eV}$, and $\left|\Delta E_{3,0}\right| = 11\,\text{eV}$. The absorbed photons have energies 6 eV, 9 eV, and 11 eV.

P23:
 Solution:

 (a) The differences in the energy levels must correspond to the energies of the emitted photons. One possible set of levels is shown in the figure below.

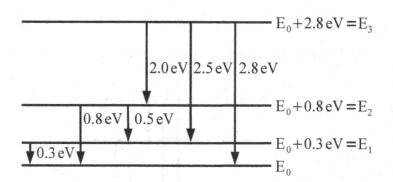

The ground state is E_0, the first excited state is $E_1 = E_0 + 0.3$ eV, the second excited state $E_2 = E_0 + 0.8$ eV, and the third excited state is $E_3 = E_0 + 2.8$ eV.

(b) No, because the energy levels are not equally spaced.

(c) Since all quantum objects are in the ground state, the only absorption lines occur for transitions from the ground state. In this case, they are: 0.3 eV, 0.8 eV, and 2.8 eV.

P27:

Solution:

(a) Treat an iron atom and bond as a quantum oscillator. Its lowest energy transition (and thus lowest energy photon emitted) is $\Delta E = \hbar\omega_0 = \hbar\sqrt{\frac{k_s}{m}}$ where k_s is the stiffness of an iron bond and m is the mass of an iron atom. The mass of an Fe atom is

$$\left(0.0558\ \frac{\text{kg}}{\text{mol}}\right)\left(\frac{1\ \text{mol}}{6.02 \times 10^{23}\ \text{atoms}}\right) = 9.3 \times 10^{-26}\ \text{kg}$$

The stiffness of a bond is found using the technique learned in Chapter 4. See the solution to 4.P.52. For iron, $k_s = 46\ \frac{\text{N}}{\text{m}}$

$$
\begin{aligned}
\omega_0 &= \sqrt{\frac{k}{m}} \\
&= \sqrt{\frac{46\ \frac{\text{N}}{\text{m}}}{9.3 \times 10^{-26}\ \text{kg}}} \\
&= 2.2 \times 10^{13}\ \frac{\text{rad}}{\text{s}}
\end{aligned}
$$

$$
\begin{aligned}
\Delta E &= \hbar\omega_0 \\
&= (6.58 \times 10^{-16}\ \text{eV}\cdot\text{s})\left(2.2 \times 10^{13}\ \frac{\text{rad}}{\text{s}}\right) \\
&= 0.015\ \text{eV}
\end{aligned}
$$

This is the lowest energy emission line and is in the IR region of the spectrum.

(b) The highest energy emission line is in the red region of the spectrum since the bar glows a dull red. If there were blue or green photons, for example, the bar would appear orange or yellow or white, a combination of the visible photons. Since red is approximately 1.8 eV, then the highest energy photons emitted are about 1.8 eV.

(c)

$$E = E_0 + N\hbar\omega$$

For large N, $E \approx N\hbar\omega$, so N is

$$\begin{aligned}
N &\approx \frac{E}{\hbar\omega} \\
&\approx \frac{1.8 \text{ eV}}{0.015 \text{ eV}} \\
&\approx 120
\end{aligned}$$

(d) Other energies are $2\hbar\omega$, $3\hbar\omega$, $4\hbar\omega$, etc. So other emitted photons have energies:

$$\begin{aligned}
2(0.015 \text{ eV}) &= 0.03 \text{ eV} \\
3(0.015 \text{ eV}) &= 0.045 \text{ eV} \\
4(0.015 \text{ eV}) &= 0.06 \text{ eV} \\
&\text{etc.}
\end{aligned}$$

CP29:
Solution:

It may help to see line numbers in the code. Here is the same program printed with line numbers.

```
from visual import *
from visual.graph import *
from random import random

Natoms = 5
# P is the probability for an atom to emit
# during a time interval dt
P = 0.1
dt=0.2 #ns
t=0
tmax = 5*dt/P # 5 mean lifetimes
# Create a bar graph
gdisplay(xtitle='t, ns',
        ytitle='Atoms in excited state')
excited = gvbars(color=color.yellow, delta=dt/2)
while t < tmax:
        rate(10)
```

```
# Show number of excited atoms remaining
excited.plot(pos=(t,Natoms))
emissions = 0
atom=0
while atom < Natoms:
        if random() < P: # emits?
                # count emissions in this dt
                emissions = emissions + 1
        atom = atom + 1
Natoms = Natoms - emissions
t = t + dt
```

(a) The command:

```
if random() < P:
```

is what determines whether an atom emits a photon or not. If a random number is less than the probability of emitting a photon, then the atom emits the photon.

(b) The mean lifetime is $dt/P = 2$ ns. Thus, 1 ns is $1/2$ of a lifetime. The probability of remaining in the same state (and NOT emitting a photon) in half of a lifetime is $e^{-1/2} = 0.61$. Therefore, the probability of emitting a photon in half of a lifetime is $1 - 0.61 = 0.39$.

(c) The count of excited atoms is decreased with the following command:

```
Natoms = Natoms - emissions
```

(d) I ran the program 10 times. Here is the time for every atom to decay in each case.

Trial	Δt for every atom to decay (s)
1	1.4
2	2.2
3	3.2
4	2.8
5	0.8
6	1.4
7	5.4
8	2.6
9	3.8
10	5.6

The longest time for these 10 trials was 5.4 s and the shortest time was 0.8 s

(e) About 750 atoms is an approximate number of atoms to get reproducible results.

(f) The time when there are 368 atoms (that have not decayed) is approximately 1.9 s. This agrees within 5% of $dt/P = 0.2/0.1 = 2$.

9   Chapter 9: Translational, Rotational, and Vibrational Energy

Q05:

Solution:

The energy equation for the point particle system differs from the energy equation for the extended system when the system is deformable, meaning that it changes shape during the interaction. They also differ when the system undergoes changes in thermal energy due to work done by friction or drag, for example.

Two such examples are:

1. A skater pushes off from a wall. (system is the skater)

2. A skydiver falls at terminal speed through the air. (system is the skydiver)

An example where the two energy equations give the same result is an orbit of a satellite orbiting a planet, where the system is the satellite. Whether you treat the satellite as a point particle or as being made of many smaller particles (the extended system), the energy equations will look the same.

P09:

Solution:

Take the origin to be at Earth's center.

$$
\begin{aligned}
\vec{r}_{CM} &= \frac{M_{Earth}\vec{R}_{Earth} + M_{Moon}\vec{R}_{Moon}}{M_{Earth} + M_{Moon}} \\
&= \frac{(6 \times 10^{24}\text{ kg})\langle 0,0,0 \rangle \text{ m} + (7 \times 10^{22}\text{ kg})\langle 4 \times 10^{8}, 0, 0 \rangle \text{ m}}{6 \times 10^{24}\text{ kg} + 7 \times 10^{22}\text{ kg}} \\
&\approx \langle 4.6 \times 10^{6}, 0, 0 \rangle \text{ m}
\end{aligned}
$$

P15:

Solution:

(a)

$$
\begin{aligned}
\vec{p}_{sys} &= \vec{p}_{1} + \vec{p}_{2} \\
&= (5\text{ kg})\langle 8, 14, 0 \rangle \text{ m/s} + (3\text{ kg})\langle -5, 9, 0 \rangle \text{ m/s} \\
&\approx \langle 25, 97, 0 \rangle \text{ kg}\cdot\text{m/s}
\end{aligned}
$$

(b)

$$
\begin{aligned}
\vec{v}_{CM} &= \frac{m_{1}\vec{v}_{1} + m_{2}\vec{v}_{2}}{m_{1} + m_{2}} \\
&= \frac{(5\text{ kg})\langle 8, 14, 0 \rangle \text{ m/s} + (3\text{ kg})\langle -5, 9, 0 \rangle \text{ m/s}}{8\text{ kg}} \\
&\approx \langle 3.125, 12.125, 0 \rangle \text{ m/s}
\end{aligned}
$$

(c)

$$K_{\text{tot}} = \frac{1}{2}m_1 \left|\vec{v}_1\right|^2 + \frac{1}{2}m_2 \left|\vec{v}_2\right|^2$$

$$= \frac{1}{2}(5 \text{ kg})(260 \text{ m}^2/\text{s}^2) + \frac{1}{2}(3 \text{ kg})(106 \text{ m}^2/\text{s}^2)$$

$$\approx 809 \text{ J}$$

(d)

$$K_{\text{trans}} = \frac{1}{2}(m_1 + m_2)\left|\vec{v}_2\right|^2$$

$$= \frac{1}{2}(8 \text{ kg})(213.03 \text{ m}^2/\text{s}^2)$$

$$\approx 627.125 \text{ J}$$

(e) $K_{\text{rel}} = K_{\text{tot}} - K_{\text{trans}} \approx 181.875 \text{ J}$

(f) For m_1:

$$\vec{v}_{\text{rel}} = \vec{v}_1 - \vec{v}_{\text{CM}} \approx \langle 4.875, 1.875, 0 \rangle \text{ m/s}$$

$$K_{\text{rel}} = \frac{1}{2}(5 \text{ kg})(27.281 \text{ m}^2/\text{s}^2)$$

$$\approx 68.203 \text{ J}$$

For m_2:

$$\vec{v}_{\text{rel}} = \vec{v}_2 - \vec{v}_{\text{CM}} \approx \langle -8.125, -3.125, 0 \rangle \text{ m/s}$$

$$K_{\text{rel}} = \frac{1}{2}(3 \text{ kg})(75.781 \text{ m}^2/\text{s}^2)$$

$$\approx 113.672 \text{ J}$$

These two values of K_{rel} should sum to the value from part (e).

P19:
  Solution:

$$I_{\text{disc}} = \frac{1}{2}MR^2$$

$$K_{\text{rot}} = \frac{1}{2}I\omega^2$$

$$= \frac{1}{2}\left(\frac{1}{2}MR^2\right)\left(\frac{2\pi}{T}\right)^2$$

$$= \frac{1}{4}(13 \text{ kg})(0.2 \text{ m})^2\left(\frac{2\pi}{0.6 \text{ s}}\right)^2$$

$$= 14.3 \text{ J}$$

P25:

   Solution:

   (a) $\left|\vec{v}_{CM}\right| = \frac{2\pi\left|\vec{r}_{ES}\right|}{T} = \frac{2\pi \times 1.5 \times 10^{11} \text{ m}}{3.15 \times 10^{7} \text{ s}} \approx 2.99 \times 10^{4} \text{ m/s}$

   (b) $K_{trans} = \frac{1}{2} M_{Earth} \left|\vec{v}_{CM}\right|^{2} = \frac{1}{2}(6 \times 10^{24} \text{ kg})(2.99 \times 10^{4} \text{ m/s})^{2} \approx 2.686 \times 10^{33} \text{ J}$

   (c) $\omega = \frac{2\pi}{T_{rot}} = \frac{2\pi}{86400 \text{ s}} \approx 7.27 \times 10^{-5} \text{ rad/s}$

   (d)

$$
\begin{aligned}
K_{rot} &= \frac{1}{2} I \omega^2 = \frac{1}{5} M_{Earth} R_{Earth}^2 \, \omega^2 \\
&= \frac{1}{5}(6 \times 10^{24} \text{ kg})(6.4 \times 10^{6} \text{ m})^2 (7.27 \times 10^{-5} \text{ rad/s})^2 \\
&\approx 2.60 \times 10^{29} \text{ J}
\end{aligned}
$$

   (e) $K_{tot} = K_{rot} + K_{trans} \approx 2.69 \times 10^{33} \text{ J}$

P29:

   Solution:

   (a)

$$
\begin{aligned}
\Delta K_{trans} &= \vec{F}_{net} \bullet \Delta \vec{r}_{CM} \\
&= F_x \Delta x_{CM} \\
&= (50 \text{ N})(1.9 \text{ m} + 1.3 \text{ m}) \\
&= (50 \text{ N})(3.2 \text{ m}) \\
&= 160 \text{ J}
\end{aligned}
$$

$$
\begin{aligned}
K_f &= 160 \text{ J} = \frac{1}{2} m v_{CM}^2 \\
v_{CM} &= \sqrt{\frac{2(160 \text{ J})}{7 \text{ kg}}} \\
&= 6.76 \text{ m/s}
\end{aligned}
$$

   (b)

$$
\begin{aligned}
\Delta E &= W \\
&= F_x \Delta x_{hand} \\
&= (50 \text{ N})(4.5 \text{ m}) \\
&= 225 \text{ J}
\end{aligned}
$$

(c)

$$
\begin{aligned}
\Delta E &= \Delta K_{trans} + \Delta E_{internal} \\
\Delta E_{internal} &= \Delta E - \Delta K_{trans} \\
&= 225 \text{ J} - 160 \text{ J} \\
&= 65 \text{ J}
\end{aligned}
$$

P35:
Solution:

(a) Apply the Energy Principle to the point-particle system.

$$
\begin{aligned}
\Delta K_{trans} &= \vec{F}_{net} \bullet \Delta \vec{r}_{CM} \\
\Delta K_{trans} &= F \Delta x_{CM} \\
\frac{1}{2}(2M)v^2_{CM,f} &= Fb \\
v_{CM,f} &= \sqrt{\frac{Fb}{M}}
\end{aligned}
$$

(b) Apply the Energy Principle to the extended system.

$$
\begin{aligned}
\Delta E &= W \\
\Delta K_{trans} + \Delta E_{int} &= F \Delta x_F \\
Fb + \Delta E_{int} &= Fd \\
\Delta E_{int} &= F(d - b)
\end{aligned}
$$

P39:
Solution:

(a) For the point particle system,

$$
\begin{aligned}
\Delta K_{trans} &= \vec{F}_{net} \bullet \Delta \vec{r}_{CM} \\
\Delta K_{trans} &= (F - 2Mg)\Delta y_{CM} \\
\Delta K_{trans} &= (F - 2Mg)(y_{CM,f} - y_{CM,i})
\end{aligned}
$$

Note that $y_{CM,f} = \frac{1}{2}(0.5 \text{ m} + 1.2 \text{ m}) = 0.85 \text{ m}$ and $y_{CM,i} = \frac{1}{2}(0.3 \text{ m} + 0.7 \text{ m}) = 0.5 \text{ m}$.

$$
\begin{aligned}
\Delta K_{trans} &= (167 \text{ N} - 2(5 \text{ kg})(9.8 \tfrac{\text{N}}{\text{kg}}))(0.85 \text{ m} - 0.5 \text{ m}) \\
&= (167 \text{ N} - 98 \text{ N})(0.35 \text{ m}) \\
&= 24.2 \text{ J}
\end{aligned}
$$

(b) For the extended system,

$$
\begin{aligned}
\Delta E &= W \\
\Delta K_{trans} + \Delta E_{vib} &= \vec{F}_{grav} \bullet \Delta r_{CM} + \vec{F}_{hand} \bullet \Delta \vec{r}_{hand} \\
\Delta K_{trans} + \Delta E_{vib} &= (-2Mg)(\Delta y_{CM}) + F\Delta y_{hand} \\
\Delta K_{trans} + \Delta E_{vib} &= -(98 \text{ N})(0.35 \text{ m}) + (167 \text{ N})(1.2 \text{ m} - 0.7 \text{ m}) \\
24.2 \text{ J} + \Delta E_{vib} &= -34.3 \text{ J} + 83.5 \text{ J} \\
\Delta E_{vib} &= 25.0 \text{ J}
\end{aligned}
$$

P43:
Solution:

(a) Apply the energy principle to the point particle system.

$$
\begin{aligned}
\Delta K_{trans} &= \int \vec{F}_{net} \cdot \Delta \vec{r}_{CM} \\
K_{f,trans} - 0 &= F\Delta x_{CM} \\
\frac{1}{2} M_{total} v_{CM,f}^2 &= Fd \\
v_{CM,f} &= \sqrt{\frac{2Fd}{M_{total}}} \\
v_{CM,f} &= \sqrt{\frac{2Fd}{M + 4m}}
\end{aligned}
$$

(b) Now apply the energy principle to the extended system.

$$
\begin{aligned}
\Delta E_{sys} &= W \\
\Delta K_{trans} + \Delta K_{rot} &= \vec{F} \cdot \vec{r}_{hand} \\
Fd + K_{rot,f} - 0 &= F\Delta x_{hand} \\
Fd + \frac{1}{2} I \omega_f^2 &= F(w + d) \\
Fd + \frac{1}{2}(4mb^2 + \frac{1}{2}MR^2)\omega_f^2 &= F(w + d) \\
\omega_f &= \left(\frac{Fw}{2mb^2 + \frac{1}{4}MR^2} \right)^{\frac{1}{2}}
\end{aligned}
$$

10   Chapter 10: Collisions

Q03:
Solution:

(B). During the time interval of the collision, $\vec{F}_{net} \approx 0$ and the momentum of the system of two asteroids is conserved.

Q11:
Solution:

The tennis ball would be more effective because it will bounce off the door, imparting a greater force on the door than would the clay, which would likely stick to the door. This follows directly from the momentum principle.

P15:
Solution:

Since the net force on the system is zero, then

$$\vec{p}_i = \vec{p}_f$$
$$\vec{p}_{1,i} + \cancel{\vec{p}_{2,i}}^{\,0} = \vec{p}_{1,f} + \vec{p}_{2,f}$$
$$m_1 v_{1,i,x} = m_1 v_{1,f,x} + m_2 v_{2,f,x}$$

It is a 1-D collision, so only the x-components of the velocity vectors are needed. Drop the x subscript for efficiency. Then,

$$m_1 v_{1,i} = m_1 v_{1,f} + m_2 v_{2,f}$$

There are 2 unknowns, the final velocities, so use the Energy Principle. It is an elastic collision.

$$\Delta K = 0$$
$$K_{1,i} + \cancel{K_{2,i}}^{\,0} = K_{1,f} + K_{2,f}$$
$$\frac{1}{2}m_1 v_{1,i}^2 = \frac{1}{2}m_1 v_{1,f}^2 + \frac{1}{2}m_2 v_{2,f}^2$$
$$m_1 v_{1,i}^2 = m_1 v_{1,f}^2 + m_2 v_{2,f}^2$$

Substitute $v_{1,f}$ from the Momentum Principle and solve for $v_{2,f}$. From the Momentum Principle,

$$v_{1,f} = v_{1,i} - \frac{m_2}{m_1}v_{2,f}$$

Substituting into the Energy Principle gives

$$v_{1,i}^2 = (v_{1,i} - \frac{m_2}{m_1}v_{2,f})^2 + \frac{m_2}{m_1}v_{2,f}^2$$

$$\cancel{v_i^2} = \cancel{v_{1,i}^2} + \left(\frac{m_2}{m_1}\right)^2 v_{2,f}^2 - 2\left(\frac{m_2}{m_1}\right)v_{1,i}v_{2,f} + \frac{m_2}{m_1}v_{2,f}^2$$

$$\left(\left(\frac{m_2}{m_1}\right)^2 + \frac{m_2}{m_1}\right)v_{2,f}^2 = 2\left(\frac{m_2}{m_1}\right)v_{1,i}v_{2,f}$$

$$\left(\frac{m_2}{m_1} + 1\right)v_{2,f} = 2v_{1,i}$$

$$v_{2,f} = \left(\frac{2}{1 + \frac{m_2}{m_1}}\right)v_{1,i}$$

Substitute this into the Momentum Principle and solve for $v_{1,f}$.

$$v_{1,f} = v_{1,i} - \frac{m_2}{m_1}v_{2,f}$$

$$= v_{1,i} - \frac{m_2}{m_1}\left(\frac{2}{1 + \frac{m_2}{m_1}}\right)v_{1,i}$$

$$= \left(1 - \frac{2}{\frac{m_1}{m_2} + 1}\right)v_{1,i}$$

$$= \left(\frac{\frac{m_1}{m_2} + 1 - 2}{\frac{m_1}{m_2} + 1}\right)v_{1,i}$$

$$v_{1,f} = \left(\frac{\frac{m_1}{m_2} - 1}{\frac{m_1}{m_2} + 1}\right)v_{1,i}$$

If $m_1 \gg m_2$, then it is like a bowling ball hitting a ping-pong ball at rest. Then, $\frac{m_2}{m_1} \approx 0$ and $\frac{m_1}{m_2} \gg 1$.

$$v_{2,f} \approx \left(\frac{2}{1+0}\right)v_{1,i} = 2v_i$$

and

$$v_{1,f} \approx \left(\frac{\frac{m_1}{m_2}}{\frac{m_1}{m_2}}\right)v_{1,i} \approx v_{1,i}$$

So in this case, the ping-pong ball has a final velocity that is twice the initial velocity of the bowling ball, and the bowling ball continues with the same velocity, as if it hit "nothing."

If $m_1 << m_2$, then it is like a ping-pong ball hitting a bowling ball. Then $\frac{m_2}{m_1} >> 1$ and $\frac{m_1}{m_2} \approx 0$.

$$v_{2,f} \approx \frac{2}{\frac{m_2}{m_1}} v_{1,i}$$

$$\approx 2\left(\frac{m_1}{m_2}\right) v_{1,i}$$

and

$$v_{1,f} \approx \left(\frac{0-1}{0+1}\right) v_{1,i} = -v_{1,i}$$

The ping-pong ball rebounds backward with the same speed and the bowling ball has a small, non-zero velocity after the collision.

P19:

Solution:

(a) Assume the net force on the system is zero. So,

$$\vec{p}_{sys,i} = \vec{p}_{sys,f}$$

$$\vec{p}_{1,i} + \vec{p}_{2,i}^{\;0} = \vec{p}_{1,f} + \vec{p}_{2,f}$$

$$(5 \text{ kg})(\langle 3300, -3100, 3400 \rangle \text{ m/s}) = (5 \text{ kg})(\langle 2800, -2400, 3700 \rangle \text{ m/s}) + \vec{p}_{2,f}$$

$$\vec{p}_{2,f} = \langle 2500, -3500, -1500 \rangle \text{ kg} \cdot \text{m/s}$$

(b)

$$K_{1,i} = \frac{1}{2} m_1 v_{1,i}^2$$

$$= \frac{1}{2}(5 \text{ kg})(5662 \text{ m/s})^2$$

$$= 8.02 \times 10^7 \text{ J}$$

(c) $K_{2,i} = \frac{1}{2} m_2 v_{2,i}^2 = 0$

(d)

$$K_{1,f} = \frac{1}{2}m_1 v_{1,f}^2$$
$$= \frac{1}{2}(5 \text{ kg})(5223 \text{ m/s})^2$$
$$= 6.82 \times 10^7 \text{ J}$$

(e) If it is an elastic collision, then $K_i = K_f$.

$$K_{1,i} + K_{2,i} = K_{1,f} + K_{2,f}$$
$$8.02 \times 10^7 \text{ J} = 6.82 \times 10^7 \text{ J} + K_{2,f}$$
$$K_{2,f} = 1.20 \times 10^7 \text{ J}$$

(f) Assume it is a closed system with $\Delta E_{sys} = 0$. Then

$$\Delta E_{sys} = 0$$
$$\Delta E_{thermal} + \Delta K = 0$$
$$\Delta E_{thermal} + (K_f - K_i) = 0$$
$$\Delta E_{thermal} + ((K_{1,f} + K_{2,f}) - (K_{1,i} + K_{2,i})) = 0$$
$$7.16 \times 10^6 \text{ J} + 6.82 \times 10^7 \text{ J} + K_{2,f} - 8.02 \times 10^7 \text{ J} = 0$$
$$K_{2,f} = 4.8 \times 10^6 \text{ J}$$

(g) During the small time interval of the collision, $Q = 0$. This was assumed when we set $\Delta E_{sys} = 0$. Note that the thermal energy of the system increased even though there was no "heat transferred" (Q) to the system.

P23:
   Solution:

(a)

$$m_{alpha} \approx 4 \text{ m}_{proton} \approx 6.68 \times 10^{-27} \text{ kg}$$
$$K = \frac{p^2}{2m}$$
$$p_{i,alpha} = \sqrt{2m_{alpha}K_{alpha}}$$
$$= \sqrt{2(6.68 \times 10^{-27} \text{ kg})(10 \times 10^6 \text{ eV})\left(\frac{1.6 \times 10^{-19} \text{ J}}{1 \text{ eV}}\right)}$$
$$= 1.46 \times 10^{-19} \text{ kg} \cdot \text{m/s}$$
$$p_{i,alpha,x} = +1.46 \times 10^{-19} \text{ kg} \cdot \text{m/s}$$

Use the results from 10.P.15. Both the Momentum Principle and Energy Principle are needed. Note that $m_{\text{gold}} \approx 197 m_{\text{proton}} = 3.29 \times 10^{-25}$ kg. Label the alpha particle 1 and the gold nucleus 2.

$$
\begin{aligned}
m_1 v_{1,f,x} &= \left(\frac{\frac{m_1}{m_2} - 1}{\frac{m_1}{m_2} + 1} \right) m_1 v_{1,i,x} \\
p_{1,f,x} &= \left(\frac{\frac{4}{197} - 1}{\frac{4}{197} + 1} \right) p_{1,i,x} \\
&= -0.96 p_{1,i,x} \\
&= -0.96 (1.46 \times 10^{-19} \text{ kg} \cdot \text{m/s}) \\
&= -1.40 \times 10^{-19} \text{ kg} \cdot \text{m/s}
\end{aligned}
$$

(b) Use the Momentum Principle. Define the system to be the alpha particle and gold nucleus. $\vec{F}_{\text{net}} = 0$ so, the momentum of the system is constant.

$$
\begin{aligned}
\vec{p}_{\text{sys,i}} &= \vec{p}_{\text{sys,f}} \\
\vec{p}_{1,i} + \cancel{\vec{p}_{2,i}}^{\,0} &= \vec{p}_{1,f} + \vec{p}_{2,f} \\
\vec{p}_{2,f} &= \vec{p}_{1,i} - \vec{p}_{1,f} \\
p_{2,f,x} &= p_{1,i,x} - p_{1,f,x} \\
&= 1.46 \times 10^{-19} \text{ kg} \cdot \text{m/s}
\end{aligned}
$$

This is nearly 2 times the initial momentum of the alpha particle, as expected from the ping-pong-ball and bowling ball example.

(c)

$$
\begin{aligned}
K_{\text{alpha,f}} &= \frac{p_{\text{alpha},f}^2}{2 m_{\text{alpha}}} \\
&= \frac{(1.4 \times 10^{-19} \text{ kg} \cdot \text{m/s})^2}{2 (6.68 \times 10^{-27} \text{ kg})} \\
&= 1.47 \times 10^{-12} \text{ J} \left(\frac{1 \text{ eV}}{1.6 \times 10^{-19} \text{ J}} \right) \\
&= 9.17 \text{ MeV}
\end{aligned}
$$

So the alpha particle lost 0.83 MeV of energy.

(d)

$$K_{gold,f} = \frac{p^2_{gold,f}}{2m_{gold}}$$

$$= \frac{(2.86 \times 10^{-19} \text{ kg} \cdot \text{m/s})^2}{2(3.29 \times 10^{-25} \text{ kg})}$$

$$= 1.24 \times 10^{-13} \text{ J} \left(\frac{1 \text{ eV}}{1.6 \times 10^{-19} \text{ J}} \right)$$

$$= 0.78 \text{ MeV}$$

(e) Apply the Energy Principle. $K_{alpha,f} = 0$ at point of closest approach. Because $m_{gold} \gg m_{alpha}$, $K_{gold} \approx 0$ at point of closest approach. It is a closed system.

$$E_i = E_f$$

$$K_i + U_{elec,i}{}^{\nearrow 0} = K_f{}^{\nearrow 0} + U_{elec,f}$$

$$K_{i,alpha} = U_{elec,f}$$

$$K_{i,alpha} = \frac{1}{4\pi\epsilon_0} \frac{q_{alpha} q_{gold}}{r_f}$$

$$(10 \text{ MeV}) \left(\frac{1.6 \times 10^{-19} \text{ J}}{1 \text{ eV}} \right) = \left(9 \times 10^9 \frac{\text{Nm}^2}{\text{C}^2} \right) \left(\frac{2(79)(1.6 \times 10^{-19} \text{ C})^2}{r_f} \right)$$

$$r_f = 2.28 \times 10^{-14} \text{ m}$$

P31:
Solution:

Initially, the Δ particle is at rest and the momentum of the system is zero. After decaying, the proton has a velocity in the $-x$ direction and the photon has a velocity in the $+x$ direction (for example) so that the total momentum of the system is also zero. (In other words, the two decay particles must travel in opposite directions in order for the momentum of the system to be conserved.)

$$E_{rest,\Delta} = 1232 \text{ MeV}$$

$$E_{rest,p} = 105.7 \text{ MeV}$$

$$E_{rest,photon} = 0 \quad \text{thus it only has kinetic energy, and } E = pc$$

$$v_{p,f} = ?$$

$$E_{photon,f} = ?$$

Define the system to be all particles and apply the Momentum Principle.

$$
\begin{aligned}
p_{ix} &= p_{fx} \\
0 &= p_{p,x} + p_{photon,x} \\
p_{p,x} &= -p_{photon,x} \\
|p_{p,x}| &= |p_{photon,x}| \\
p_p &= \frac{E_{photon}}{c} \\
\gamma_f m v &= \frac{E_{photon}}{c}
\end{aligned}
$$

Apply the Energy Principle to the system.

$$
\begin{aligned}
E_i &= E_f \\
E_{rest,\Delta} &= E_p + E_{photon} \\
E_{rest,\Delta} &= \gamma E_{rest,p} + E_{photon}
\end{aligned}
$$

We now have two equations and two unknowns. First, solve for E_{photon} in the Momentum Principle.

$$
\begin{aligned}
\gamma_f m v &= \frac{E_{photon}}{c} \\
E_{photon} &= \gamma_f m v c \\
E_{photon} &= \gamma_f m c^2 \left(\frac{v}{c}\right) \\
E_{photon} &= \gamma_f E_{rest,p} \left(\frac{v}{c}\right)
\end{aligned}
$$

Now substitute into the Energy Principle and solve for v, the speed of the proton after the decay. Note that $\gamma = 1/\sqrt{1 - (v/c)^2}$.

$$
\begin{aligned}
E_{rest,\Delta} &= \gamma E_{rest,p} + E_{photon} \\
&= \gamma_f E_{rest,p} + \gamma_f E_{rest,p}\frac{v}{c} \\
&= \gamma_f E_{rest,p} \left(1 + \frac{v}{c}\right) \\
\frac{E_{rest,\Delta}}{E_{rest,p}} &= \frac{1 + \frac{v}{c}}{\sqrt{1 - \left(\frac{v}{c}\right)^2}}
\end{aligned}
$$

For the purpose of doing the algebra, it is convenient to set $\beta = v/c$ and $\alpha = E_{rest,\Delta}/E_{rest,p} = 1.31$. Of course, you can always use a calculator or computer algebra system to solve for v.

$$\alpha = \frac{1+\beta}{\sqrt{1-(\beta^2)}}$$

$$\alpha^2 = \frac{(1+\beta)^2}{1-\beta^2}$$

$$= \frac{(1+\beta)(1+\beta)}{(1-\beta)(1+\beta)}$$

$$= \frac{(1+\beta)}{(1-\beta)}$$

$$\alpha^2 - \beta\alpha^2 = 1+\beta$$

$$\beta(\alpha^2+1) = \alpha^2-1$$

$$\beta = \frac{\alpha^2-1}{(\alpha^2+1)}$$

$$= \frac{1.31^2-1}{(1.31^2+1)}$$

$$= 0.266$$

$$v = 0.266c$$

$$= 8\times10^7 \text{ m/s}$$

Calculate γ_f for the proton after the decay.

$$\gamma = \frac{1}{\sqrt{1-(v/c)^2}}$$

$$= \frac{1}{\sqrt{1-0.266^2}}$$

$$= 1.037$$

Now calculate the energy of the photon using the equation derived from the Momentum Principle.

$$E_{photon} = \gamma_f E_{rest,p}\left(\frac{v}{c}\right)$$

$$= (1.037)(938 \text{ MeV})(0.266)$$

$$= 259 \text{ MeV}$$

CP35:

Solution:

Here is a program that plots the x-momentum of the alpha particle and gold nucleus and the y-momentum of the alpha particle and gold nucleus. On the same graphs, it also plots the total x-momentum and total y-momentum of the system.

```
from visual import *
from visual.graph import *
scene.width = 1024
scene.height = 600
q_e = 1.6e-19
m_p = 1.7e-27
oofpez = 9e9
```

```
m_Au = (79+118) * m_p
m_Alpha = (2+2) * m_p
qAu = 2 * q_e
qAlpha = 79 * q_e
deltat = 1e-23
Au = sphere(pos=vector(0,0,0), radius=4e-15,
            color=color.yellow, make_trail=True, opacity=0.7)
Alpha = sphere(pos=vector(-1e-13,5e-15,0),
               radius=1e-15, color=color.magenta,
               make_trail=True)
p_Au = m_Au*vector(0,0,0)
p_Alpha = vector(1.043e-19,0,0)
t=0

##x component graphs
gdx = gdisplay(x=0,y=450, width=500, title='p_x',
               xtitle='t (s)',ytitle='p_x (kg m/s)')
p_Au_x_graph = gcurve(color=color.yellow)
p_Alpha_x_graph = gcurve(color=color.magenta)
psys_x_graph = gcurve(color=color.cyan)

## y component graphs
gdy = gdisplay(x=500,y=450,width=500,title='p_y',
               xtitle='t (s)',ytitle='p_y (kg m/s)')
p_Au_y_graph = gcurve(color=color.yellow)
p_Alpha_y_graph = gcurve(color=color.magenta)
psys_y_graph = gcurve(color=color.cyan)


while t<1.3e-20:
    rate(100)

    r = Alpha.pos - Au.pos
    rmag=mag(r)
    runit=norm(r)
    Fonalpha=oofpez*qAu*qAlpha/rmag**2 * runit
    Fongold=-Fonalpha

    p_Alpha = p_Alpha + Fonalpha*deltat
    Alpha.pos = Alpha.pos+(p_Alpha/m_Alpha) * deltat

    p_Au = p_Au + Fongold*deltat
    Au.pos = Au.pos+(p_Au/m_Au) * deltat

    # update graphs
    p_Au_x_graph.plot(pos=(t,p_Au.x))
    p_Alpha_x_graph.plot(pos=(t,p_Alpha.x))
    psys_x_graph.plot(pos=(t,p_Au.x+p_Alpha.x))
    p_Au_y_graph.plot(pos=(t,p_Au.y))
    p_Alpha_y_graph.plot(pos=(t,p_Alpha.y))
    psys_y_graph.plot(pos=(t,p_Au.y+p_Alpha.y))
```

```
t = t + deltat
```

The graphs created by the program are shown below.

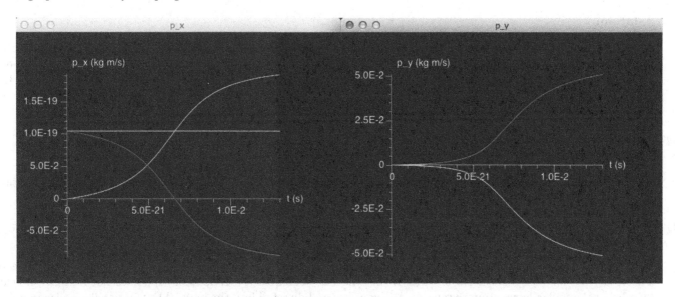

The graphs show that p_x and p_y for the system are constant; therefore, the momentum of the system is conserved. If you find that momentum is not conserved, then it's possible that the time step Δt is too large. The momentum update equation and the position update equation are approximations that are only value for very small Δt.

11   Chapter 11: Angular Momentum

Q03:
Solution:

$\vec{L}$ is constant if the net torque on the system is zero. Of the net torque is zero in the x-direction and non-zero in the y-direction, then the y-component of $\vec{L}$ will change. Suppose there is a wrench that is rotated in the x-z plane with an increasing angular speed. Then there is a non-zero torque in the y-direction and L_y changes though L_x and L_z are constant.

Q09:
Solution:

If it's far from other objects, then there is no torque exerted on the star, so its angular momentum remains constant.

Q11:
Solution:

If the stripe remains vertical, then translational angular momentum relative to A is nonzero, directed into the page. The rotational angular momentum is zero. If the wheel is welded to the rod, the stripe does not remain vertical the translational angular momentum relative to A is nonzero, and directed into the page. The rotational angular momentum is nonzero, also directed into the page.

P15:
Solution:

$$
\begin{aligned}
\vec{L}_A &= \vec{r}_A \times \vec{p} \\
&= <6,6,0> \times <-11,13,0> \\
&= \langle 0,0,144 \rangle \ \text{kg}\frac{\text{m}^2}{\text{s}}
\end{aligned}
$$

P19:
Solution:

$$
\begin{aligned}
\left| \vec{L}_{rot} \right| &= I\,|\vec{\omega}| \\
&= \frac{2}{5} M_{Earth} R_{Earth}^2 \frac{2\pi}{86400 \ \text{s}} \\
&\approx \frac{2}{5}\left(6 \times 10^{24} \ \text{kg}\right)\left(6.4 \times 10^6 \ \text{m}\right)\frac{2\pi}{86400 \ \text{s}} \\
&\approx 7.1 \times 10^{33} \ \text{kg} \cdot \text{m}^2/\text{s}
\end{aligned}
$$

P25:
Solution:

(a)

$$
\begin{aligned}
I &= I_1 + I_2 \\
I &= 2mr^2 \\
&= 2(1.7\ \text{kg})(0.15\ \text{m})^2 \\
&= 0.0765\ \text{kg} \cdot \text{m}^2
\end{aligned}
$$

(b)

$$
\begin{aligned}
\left|\vec{L}_{rot}\right| &= I\omega \\
&= (0.0765\ \text{kg} \cdot \text{m}^2)\left(\frac{2\pi}{0.5\ \text{s}}\right) \\
&= 0.91\ \text{kg} \cdot \frac{\text{m}^2}{\text{s}}
\end{aligned}
$$

(c)

$$
\begin{aligned}
K_{rot} &= \frac{L_{rot}^2}{2I} \\
&= \frac{(0.961\ \text{kg} \cdot \frac{\text{m}^2}{\text{s}})}{2(0.0765\ \text{kg} \cdot \text{m}^2)} \\
&= 6.28\ \text{J}
\end{aligned}
$$

P27:
   Solution:

(a)

$$
\begin{aligned}
\vec{p}_{total} &= m_1\vec{v}_1 + m_2\vec{v}_2 = m\left(\vec{v}_1 + \vec{v}_2\right) \\
&\approx \langle 29.4, 0, 0 \rangle\ \text{kg} \cdot \text{m/s}
\end{aligned}
$$

(b)

$$
\begin{aligned}
\vec{v}_{cm} &= \frac{\vec{p}_{total}}{2m} \\
&\approx \langle 49, 0, 0 \rangle\ \text{m/s}
\end{aligned}
$$

(c)

$$
\begin{aligned}
\vec{L}_{A,total} &= \vec{r}_1 \times \vec{p}_1 + \vec{r}_2 \times \vec{p}_2 \\
&\approx \langle 0, 0, -25.9 \rangle\ \text{kg} \cdot \text{m}^2/\text{s}
\end{aligned}
$$

(d)

$$
\begin{aligned}
\vec{L}_{rot} &= \vec{r}_{1,cm} \times \vec{p}_1 + \vec{r}_{2,cm} \times \vec{p}_2 \\
&\approx \langle 0, 0, 4.95 \rangle\ \text{kg} \cdot \text{m}^2/\text{s}
\end{aligned}
$$

(e)

$$\vec{L}_{trans,A} = \vec{r}_{cm,A} \times \vec{p}_{total}$$
$$\approx \langle 0, 0, -30.9 \rangle \ \mathrm{kg \cdot m^2/s}$$

Check: $\vec{L}_{A,total} = \vec{L}_{trans,A} + \vec{L}_{rot}$

(f)

$$\vec{p}_f = \vec{p}_i + \vec{F}_{net}\Delta t$$
$$\approx \langle 32.9, 0, 0 \rangle \ \mathrm{kg \cdot m/s}$$

P31:
Solution:

$$\vec{L}_{A,f} = \vec{L}_{A,i} + \vec{\tau}_A \Delta t$$
$$\approx \langle 3, 5, -2 \rangle \ \mathrm{kg \cdot m^2/s} + \langle 10, -12, 20 \rangle \ \mathrm{N \cdot m}\,(0.1 \ \mathrm{s})$$
$$\approx \langle 4, 3.8, 0 \rangle \ \mathrm{kg \cdot m^2/s}$$

P35:
Solution:

(a)

$$\vec{L}_{C,i} = \vec{L}_{C,wheel,i} + \vec{L}_{C,clay,i}$$
$$L_{c,i,z} = -I\omega_i + \left|\vec{r}_C\right|\left|\vec{p}_i\right|\sin\theta$$
$$= -(\frac{1}{2}MR^2)\omega_i + m\left|\vec{v}_i\right|R\sin 45°$$
$$= -\frac{1}{2}(4.8 \ \mathrm{kg})(0.9 \ \mathrm{m})^2(0.33 \ \mathrm{rad/s}) + (0.5 \ \mathrm{kg})(0.9 \ \mathrm{m})(\frac{\sqrt{2}}{2})$$
$$= -0.642 \ \mathrm{kg} \cdot \frac{\mathrm{m}^2}{\mathrm{s}} + 0.318 \ \mathrm{kg} \cdot \frac{\mathrm{m}^2}{\mathrm{s}}$$
$$= -0.323 \ \mathrm{kg} \cdot \frac{\mathrm{m}^2}{\mathrm{s}}$$

(b) The support exerts a force on the axle, but this force exerts no torque about C on the system. The gravitational force on the wheel exerts no torque about C on the system. The only torque on the system is the gravitational force on m. However, during the small time interval of the collision, $\vec{L}_C$ is approximately constant. Thus,

$$\vec{L}_{C,f} = \vec{L}_{C,i}$$
$$\vec{L}_{C,f} = -0.323 \ \mathrm{kg} \cdot \frac{\mathrm{m}^2}{\mathrm{s}}$$

(c)

$$\begin{aligned}
\vec{L}_{C,f} &= \vec{L}_{C,f,wheel} + \vec{L}_{C,f,clay} \\
L_{C,z,f} &= I_{wheel}\,\omega_f + I_{clay}\,\omega_f \\
&= \frac{1}{2}MR^2\omega_f + mR^2\omega_f \\
&= (\frac{1}{2}MR^2 + mR^2)\omega_f \\
\omega_f &= \frac{-0.323\ \text{kg}\cdot\frac{m^2}{s}}{\frac{1}{2}(4.8\ \text{kg})(0.9\ \text{m})^2 + (0.5\ \text{kg})(0.9\ \text{m})^2} \\
&= -0.138\ \text{rad/s}
\end{aligned}$$

(d) Before the collision, $\vec{p}_{i,sys} = \vec{p}_{i,clay}$ which is downward. After the collision, $\vec{p}_{f,sys} = \vec{p}_{f,clay}$ which is downward and to the left and its y-component is less than $\vec{p}_{i,clay}$. Thus, the support exerts a force that is upward and to the left on the system. Thus, the answer is (1).

P41:
Solution:

(a) Apply the Momentum Principle to the apparatus. It is in equilibrium, $\vec{F}_{net} = 0$. Thus, the upward force of tension on the rod is equal in magnitude to the gravitational force by Earth on the system. So, $\vec{F}_{T,string} = (m_1 + m_2)g = (0.484\ \text{kg} + 0.273\ \text{kg})(9.8\ \text{N/kg}) = 7.42\ \text{N}$.

(b) Call the mass on the left end m_1 and the other mass m_2. Define the origin $x = 0$ to be the left end of the rod. In order to be balanced, the net torque on the system must be zero. As a result, the x-component of the center of mass of the system must be at the location of the string. The x-component of the center of mass is

$$\begin{aligned}
x_{CM} &= \frac{m_1 x_1 + m_2 x_2}{m_1 + m_2} \\
&= \frac{0 + (0.273\ \text{kg})(0.49\ \text{m})}{(0.484\ \text{kg} + 0.273\ \text{kg})} \\
&= 0.177\ \text{m}
\end{aligned}$$

This is closer to the left end of the rod, which makes sense because the mass at the left end is greater than the mass at the right end. If they were equal masses, then the center of mass would be in the geometric center of the rod.

P47:
Solution:

Apply the Momentum Principle. Treat the barbell as a particle located at its center of mass with total mass $2m_1$. Since the net force on the barbell/m_2 system is zero, then the linear momentum is conserved. Define +x to the right, +y to the top of the page, and +z outward.

x:

$$
\begin{aligned}
\vec{p}_{sys,i} &= \vec{p}_{sys,f} \\
p_{1,i,x} + p_{2,i,x} &= p_{1,f,x} + p_{2,f,x} \\
2m_1 v_1 + -m_2 v_2 \cos\theta_2 &= m_1 v_3 \cos\theta_3 + m_2 v_4 \cos\theta_4
\end{aligned}
$$

y:

$$
\begin{aligned}
\cancel{p_{1,i,y}}^{0} + p_{2,i,y} &= p_{1,f,y} + p_{2,f,y} \\
-m_2 v_2 \sin\theta_2 &= 2m_1 v_3 \sin\theta_3 - m_2 v_4 \sin\theta_4
\end{aligned}
$$

Apply the Momentum Principle to the system of barbell $+ m_2$. Since the net torque on the system is zero, then $\vec{L}_{sys}$ is conserved. $\vec{L}$ is in the z-direction since the motion is in the x,y plane. Calculate $\vec{L}$ about the center of the barbell.

z:

$$
\begin{aligned}
L_{1,i,z} + L_{2,i,z} &= L_{1,f,z} + L_{2,f,z} \\
\cancel{L_{1,trans,i,z}}^{0} + L_{1,rot,i,z} + (\vec{r}_2 \times \vec{p}_2)_z &= \cancel{L_{1,trans,f,z}}^{0} + L_{1,rot,f,z} + (\vec{r}_2 \times \vec{p}_4)_z
\end{aligned}
$$

Note that $\vec{L}_{trans}$ barbell about its center is zero since $\vec{v}$ is the center-of-mass velocity.

$$
\begin{aligned}
I_{1,z}\omega_{1,i,z} + (<0,-b,0> \times <-m_2 v_2 \cos\theta_2, -m_2 v_2 \sin\theta_2, 0>)_z &= I_{1,f}\omega_{1,f,z} + (<0,-b,0> \times <+m_2 v_4 \cos\theta_4, -m_2 v_4 \sin\theta_4 \\
2m_1 \left(\frac{L}{2}\right)^2 \omega_1 + -bm_2 v_2 \cos\theta_2 &= 2m_1 \left(\frac{L}{2}\right)^2 \omega_2 + bm_2 v_4 \cos\theta_4
\end{aligned}
$$

We now have the following three equations. Given all other quantities, the unknowns v_3, θ_3, and ω_2 can be found.

$$
\begin{aligned}
2m_1 v_1 + -m_2 v_2 \cos\theta_2 &= m_1 v_3 \cos\theta_3 + m_2 v_4 \cos\theta_4 \\
-m_2 v_2 \sin\theta_2 &= 2m_1 v_3 \sin\theta_3 - m_2 v_4 \sin\theta_4 \\
2m_1 \left(\frac{L}{2}\right)^2 \omega_1 + -bm_2 v_2 \cos\theta_2 &= 2m_1 \left(\frac{L}{2}\right)^2 \omega_2 + bm_2 v_4 \cos\theta_4
\end{aligned}
$$

P51:
  Solution:

  (a) Apply the Momentum Principle. Define the system to be the rod and the meteorite. Since $\vec{F}_{net} = 0$, $\vec{p}_{sys}$ is constant.

$$
\vec{p}_{sys,i} = \vec{p}_{sys,f}
$$

x:

$$mv_i \cos\theta_i \;=\; -mv_f \cos\theta_f + Mv_x$$

y:

$$mv_i \sin\theta_i \;=\; -mv_f \sin\theta_f + Mv_y$$

so

$$v_x \;=\; \frac{m(v_i \cos\theta_i + v_f \cos\theta_f)}{M}$$

and

$$v_y \;=\; \frac{m(v_i \sin\theta_i + v_f \sin\theta_f)}{M}$$

Substituting known values gives

$$\begin{aligned} v_x &= \frac{m(v_i \cos\theta_i + v_f \cos\theta_f)}{M} \\ &= \frac{(0.06\text{ kg})((200\text{ m/s})\cos 26^\circ + (60\text{ m/s})\cos 82^\circ)}{1.3\text{ kg}} \\ &= 8.68\text{ m/s} \end{aligned}$$

$$\begin{aligned} v_y &= \frac{m(v_i \sin\theta_i + v_f \sin\theta_f)}{M} \\ &= \frac{(0.06\text{ kg})((200\text{ m/s})\sin 26^\circ + (60\text{ m/s})\sin 82^\circ)}{1.3\text{ kg}} \\ &= 1.3\text{ m/s} \end{aligned}$$

(b) Apply the Angular Momentum Principle about the center of the stick. Label the center of the stick as point C. Since $\vec{\tau}_{net,C} = 0$ for the system, then $\vec{L}_C$ is constant. Calculate $\vec{L}$ about the center of the stick. The $\perp$ distance from the CM of the stick to the dotted line through the initial velocity vector is $r_\perp = d\cos\theta_i$. The $\perp$ distance for v_f is $r_\perp = d\cos\theta_f$

$$\vec{L}_{sys,i} = \vec{L}_{sys,f}$$
$$L_{m,trans,i,z} = L_{m,trans,f,z} + I\omega_{f,z}$$
$$-r_{\perp,i}mv_i = r_{\perp,f}mv_f + I\omega_{f,z}$$
$$-d\cos\theta_i\, mv_i - d\cos\theta_f\, mv_f = \frac{1}{12}ML^2\omega_{f,z}$$
$$\omega_{f,z} = \frac{-md(v_i\cos\theta_i + v_f\cos\theta_f)}{\frac{1}{12}ML^2}$$
$$|\vec{\omega}| = \frac{md(v_i\cos\theta_i + v_f\cos\theta_f)}{\frac{1}{12}ML^2}$$

and the direction of $\vec{\omega}$ is in the -z direction, $< 0, 0, -1 >$. Substituting known values gives:

$$\omega_z = \frac{-md(v_i\cos\theta_i + v_f\cos\theta_f)}{\frac{1}{12}mL^2}$$
$$= \frac{-(0.6\text{ kg})(0.2\text{ m})((200\text{ m/s})\cos 26° + (60\text{ m/s})\sin 80°)}{\frac{1}{12}(1.3\text{ kg})(0.4\text{ kg})^2}$$
$$= -166\text{ rad/s} \approx -26\text{ rev/s}$$

The angular velocity can be expressed $\vec{\omega} = < 0, 0, -166 >$ rad/s.

(c) It is a closed system, so $\Delta E_{sys} = 0$.

$$\Delta E_{sys} = 0$$
$$\Delta K + \Delta E_{internal} = 0$$
$$\Delta E_{internal} = -\Delta K$$
$$= -(\Delta K_{trans} + \Delta K_{rot})$$
$$= K_{i,trans} - K_{f,trans} + K_{i,rot} - K_{f,rot}$$

$$K_{m,i} = \frac{1}{2}mv_i^2$$
$$K_{m,f} = \frac{1}{2}mv_f^2$$
$$K_{trans,stick,f} = \frac{1}{2}Mv^2$$
$$K_{rot,stick,f} = \frac{1}{2}I\omega^2$$
$$= \frac{1}{2}(\frac{1}{12}ML^2)\omega^2$$
$$= \frac{1}{24}ML^2\omega^2$$

Thus, $\Delta E_{int} = \frac{1}{2}m(v_i^2 - v_f^2) - \frac{1}{2}Mv^2 - \frac{1}{24}ML^2\omega^2$ where $v = v_x^2 + v_y^2$.

Substituting known values gives

$$
\begin{aligned}
\Delta E_{int} &= \frac{1}{2}m(v_i^2 - v_f^2) - \frac{1}{24}ML^2\omega^2 \\
&= \frac{1}{2}(0.06 \text{ kg})(200^2 - 60^2) - \frac{1}{2}(1.3 \text{ kg})(8.68^2 + 1.3^2) - \frac{1}{24}(1.3 \text{ kg})(0.4 \text{ m})^2(166 \text{ rad/s})^2 \\
&= 1092 \text{ J} - 50.1 \text{ J} - 239 \text{ J} \\
&= 803 \text{ J}
\end{aligned}
$$

The internal energy increases about 800 J as a result of the interaction.

P55:
  Solution:

   (a)

$$
\begin{aligned}
\vec{L}_{rot} &= I\vec{\omega} \\
\vec{L}_{rot} &= \frac{1}{2}MR^2\vec{\omega} \\
&= \frac{1}{2}(0.4 \text{ kg})(0.08 \text{ m})^2\langle 0, 0, -20 \rangle \text{ rad/s} \\
&= \langle 0, 0, -0.0256 \rangle \text{ kg} \cdot \frac{m^2}{s}
\end{aligned}
$$

   (b)

$$
\begin{aligned}
\vec{\tau}_{CM} &= \frac{\Delta\vec{L}_{rot}}{\Delta t} \\
\tau_{CM,z} &= \frac{\Delta L_{rot,z}}{\Delta t} \\
L_{z,f} &= L_{z,f} + \tau_z\Delta t \\
&= -0.0256 \text{ kg} \cdot \frac{m^2}{s} - (0.8 \text{ N}\cdot\text{m})(0.2) \\
&= -0.186 \text{ kg} \cdot \frac{m^2}{s}
\end{aligned}
$$

   (c)

$$
\begin{aligned}
L_{z,f} &= I\omega_{z,f} \\
\omega_{z,f} &= \frac{-0.186 \text{ kg} \cdot \frac{m^2}{s}}{\frac{1}{2}(0.4 \text{ kg})(0.08 \text{ m})^2} = -145 \text{ rad/s}
\end{aligned}
$$

Note that it sped up by a factor of 7 during the time interval of 0.2 s

P59:

Solution:

Apply the Angular Momentum Principle to the disk about its axle. Calculate torque about the axle. The only force that exerts a torque is the force by the string which is tangent to the disk.

$$
\begin{aligned}
\vec{\tau}_{net} &= \frac{\Delta \vec{L}}{\Delta t} \\
\tau_z &= \frac{\Delta L_z}{\Delta t} \\
-F_{string} R &= \frac{I \Delta \omega_z}{\Delta t} \\
-(25 \text{ N})(0.2 \text{ m}) &= \frac{(1.5 \text{ kg} \cdot \text{m}^2)(\omega_{z,f} - (-2 \text{ rad/s}))}{0.1 \text{ s}} \\
\omega_{z,f} &= -3.67 \text{ rad/s}
\end{aligned}
$$

It is rotating faster in the -z direction.

P63:

Solution:

Apply the Angular Momentum Principle to the apparatus, about its CM.

$$
\begin{aligned}
\tau_{about\ CM,z} &= \frac{\Delta L_z}{\Delta t} \\
-FR &= I \frac{\Delta \omega_z}{\Delta t} \\
\Delta \omega_z &= \frac{-FR\Delta t}{I} \\
&= \frac{-FR\Delta t}{(4mb^2 + \frac{1}{2}MR^2)}
\end{aligned}
$$

Note that $\Delta \omega_z$ is negative, meaning that ω_z increases in the -z direction. Since $\omega_i = 0$, then $\omega_f = \Delta \omega$.

$$
\begin{aligned}
\omega_z &= \frac{-FR\Delta t}{(4mb^2 + \frac{1}{2}MR^2)} \\
&= \frac{-(21 \text{ N})(0.11 \text{ m})(0.2 \text{ s})}{4(0.4 \text{ kg})(0.14 \text{ m})^2 + \frac{1}{2}(1.2 \text{ kg})(0.11 \text{ m})^2} \\
&= -12 \text{ rad/s}
\end{aligned}
$$

The angular speed is $\omega = 12$ rad/s.

P67:

   Solution:

$$\begin{aligned} \omega_i &= 24 \text{ rad/s} \\ \omega_f &= 36 \text{ rad/s} \\ \Delta t &= 3 \text{ s} \end{aligned}$$

(a)

$$\alpha = \frac{\Delta\omega}{\Delta t} = \frac{36 \text{ rad/s} - 24 \text{ rad/s}}{3 \text{ s}} = 4 \text{ rad/s}^2$$

(b)

$$\omega_{ave} = \frac{\omega_i + \omega_f}{2} = \frac{24 \text{ rad/s} + 36 \text{ rad/s}}{2} = 30 \text{ rad/s}$$

(c)

$$\Delta\theta = \omega_{ave}\Delta t = (30 \text{ rad/s})(3 \text{ s}) = 90 \text{ rad}$$

(d) In degrees,

$$\Delta\theta = (90 \text{ rad})\left(\frac{180°}{\pi \text{ rad}}\right) = 5160°$$

P73:

   Solution:

First sketch a picture of the situation that shows the forces acting on each ball. Let's place the solid ball on the left and the hollow ball on the right. Call the solid ball #1 and the hollow ball #2. Define $x = 0$ to be the initial location of the solid ball and $x = L$ to be the initial location of the hollow ball.

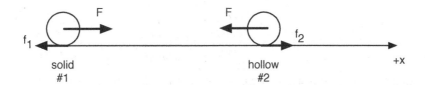

We need to think carefully about the problem before we start the solution. Here are a few things to notice:

- If the balls were in space or were on a frictionless surface, then they would meet in the middle (at $L/2$) at the center of mass of the system. However, they are on a surface with friction, and it's the friction force that causes them to roll.

- The friction force on each ball is not the same. We know this because of an analogous situation that you might be familiar with. Suppose that you release the two balls at the top of a ramp, and they roll to the bottom. The gravitational force on each ball, parallel to the ramp, is the same (just like this situation). The solid ball gets to the bottom of the ramp first and thus has a greater center of mass acceleration than the hollow ball. This means that the frictional force on the hollow ball is greater than the solid ball. If we apply the same reasoning to this situation, we expect the solid ball to have a greater center of mass acceleration than the hollow ball. Thus they should meet somewhere to the right at $x > L/2$.

- Since $f_2 > f_1$, then there is a net external force on the system, and the center of mass of the system (of balls) accelerates to the right.

- F is an electric force with magnitude

$$F = \frac{1}{4\pi\varepsilon_o} \frac{|q_1 q_2|}{r^2}$$

where q_1 and q_2 are the charges on each ball. The charges must have opposite signs in order for the force to be attractive. This force increases as the distance between the balls r decreases. Thus the acceleration of each ball will not be constant (unless f were to compensate just the right amount to keep a constant, but that's probably not the case). Therefore we cannot make the simplifying assumption that the acceleration of each ball is constant.

- The moment of inertia of the solid ball about its center is $I_1 = 2/5 mR^2$. The moment of inertial of the hollow ball about its center is $I_2 = 2/3 m * R^2$.

- The balls roll without slipping. Thus $v_{cm,ball} = \omega R$ and $a_{cm,ball} = \alpha R$.

Let's begin by analyzing ball 1. Its x-acceleration (of its CM) is to the right, yet its angular acceleration is in the $-z$ direction. Written in component form, $a_{1,x} = -\alpha_1 R$ where α_1 is now the z-component of the angular acceleration vector.

Apply the Angular Momentum Principle about the center of ball 1 in the z-direction. The frictional force f_1 exerts a torque in the $-z$ direction. Solve for the frictional force.

$$
\begin{aligned}
\tau_{net,C,z} &= \frac{\Delta L_{z,1}}{\Delta t} \\
-f_1 R &= I_1 \alpha_1 \\
f_1 &= \frac{-I_1}{R}\alpha_1
\end{aligned}
$$

Apply the Momentum Principle in the x-direction to the CM of ball 1. Write $dp_x/dt = ma_x$. Note that $a_x = +a_1$ where a_1 is the magnitude of the ball's acceleration.

$$
\begin{aligned}
F_{net,x} &= \frac{dp_x}{dt} \\
F_{net,x} &= ma_x \\
F - f_1 &= ma_1
\end{aligned}
$$

Substitute the frictional force from the Angular Momentum Principle.

$$
\begin{aligned}
F - f_1 &= ma_1 \\
F - \frac{-I_1}{R}\alpha_1 &= ma_1 \\
F + \frac{I_1}{R}\alpha_1 &= ma_1
\end{aligned}
$$

Solve $a_{1,x} = -\alpha_1 R$ for α_1 and substitute. Then solve for F. Also substitute I_1 for a solid ball.

$$
\begin{aligned}
F + \frac{I_1}{R}\left(\frac{-a_1}{R}\right) &= ma_1 \\
F &= ma_1 + \frac{I_1}{R^2}a_1 \\
&= \left(m + \frac{I_1}{R^2}\right)a_1 \\
&= \left(m + \frac{\frac{2}{5}mR^2}{R^2}\right)a_1 \\
&= \left(m + \frac{2}{5}m\right)a_1 \\
&= \left(1 + \frac{2}{5}\right)ma_1 \\
&= \frac{7}{5}ma_1
\end{aligned}
$$

Note that we can now think of ball 1 as a point particle in space with an effective mass of

$$
\begin{aligned}
m_{\text{eff},1} &= \left(m + \frac{I_1}{R^2}\right) \\
&= \frac{7}{5}m
\end{aligned}
$$

Then we can apply the Momentum Principle to point particle 1 using

$$
\begin{aligned}
F_{\text{net},1,x} &= m_{\text{eff}}\, a_{1,x} \\
F &= m_{\text{eff},1}\, a_{1,x}
\end{aligned}
$$

In other words, the effective mass already accounts for the fact that there is friction, so we can model the ball as a point particle in space with no friction.

Now we have to repeat this analysis for ball 2 so that we can see how their accelerations are related. We expect their accelerations to be different.

The x-acceleration of ball 2 (of its CM) is to the left, yet its angular acceleration is in the $+z$ direction. Written in component form, $a_{2,x} = -\alpha_2 R$ where α_2 is now the z-component of the angular acceleration vector.

Apply the Angular Momentum Principle to the center of ball 2 in the z-direction. The frictional force f_2 exerts a torque in the $+z$ direction. Solve for the frictional force.

$$
\begin{aligned}
\tau_{net,C,z} &= \frac{\Delta L_{z,2}}{\Delta t} \\
f_2 R &= I_2 \alpha_2 \\
f_2 &= \frac{I_2}{R}\alpha_2
\end{aligned}
$$

Apply the Momentum Principle in the x-direction to the CM of ball 2. Write $dp_x/dt = ma_x$. Note that $a_x = -a_2$ where a_2 is the magnitude of the ball's acceleration.

$$
\begin{aligned}
F_{\text{net,x}} &= \frac{dp_x}{dt} \\
F_{\text{net,x}} &= ma_{2,x} \\
f_2 - F &= -ma_2
\end{aligned}
$$

Substitute the frictional force from the Angular Momentum Principle.

$$
\begin{aligned}
f_2 - F &= -ma_2 \\
\frac{I_2}{R}\alpha_2 - F &= -ma_2 \\
F - \frac{I_2}{R}\alpha_2 &= ma_2
\end{aligned}
$$

Solve $a_{2,x} = -\alpha_2 R$ for α_2 and substitute. Then solve for F. Also substitute I_2 for a solid ball.

$$
\begin{aligned}
F - \frac{I_2}{R}\left(\frac{-a_2}{R}\right) &= ma_2 \\
F &= ma_2 + \frac{I_2}{R^2}a_2 \\
&= \left(m + \frac{I_2}{R^2}\right)a_2 \\
&= \left(m + \frac{\frac{2}{3}mR^2}{R^2}\right)a_2 \\
&= \left(m + \frac{2}{3}m\right)a_2 \\
&= \left(1 + \frac{2}{3}\right)ma_2 \\
&= \frac{5}{3}ma_2
\end{aligned}
$$

Note that we can now think of ball 2 as a point particle in space with an effective mass of

$$
\begin{aligned}
m_{\text{eff,2}} &= \left(m + \frac{I_2}{R^2}\right) \\
&= \frac{5}{3}m
\end{aligned}
$$

Then we can apply the Momentum Principle to point particle 2 using

$$
\begin{aligned}
F_{\text{net,2,x}} &= m_{\text{eff,2}}\,a_{2,x} \\
F &= m_{\text{eff,2}}\,a_{2,x}
\end{aligned}
$$

Now that we have expressions for the magnitude of the acceleration of each ball, we can solve for the ratio.

$$
\begin{aligned}
\frac{a_1}{a_2} &= \frac{\frac{F}{\frac{7}{5}m}}{\frac{F}{\frac{5}{3}m}} \\
&= \frac{\frac{5}{3}}{\frac{7}{5}} \\
&= \left(\frac{5}{3}\right)\left(\frac{5}{7}\right) \\
&= \frac{25}{21} \\
&= 1.19
\end{aligned}
$$

As expected, the solid ball has a greater acceleration. In fact it is 1.19 times the acceleration of the hollow ball. If we model the two balls as point particles in space with different effective masses, then they will meet at the location of center of mass of the system. Let's find the center of mass of the system.

$$
\begin{aligned}
x_{\text{CM}} &= \frac{m_1 x_1 + m_2 x_2}{m_1 + m_2} \\
&= \frac{0 + m_2 L}{m_1 + m_2} \\
&= \frac{\frac{5}{3}m}{\frac{7}{5}m + \frac{5}{3}m}L \\
&= \frac{\frac{5}{3}}{\frac{46}{15}}L \\
&= \frac{5}{3}\frac{15}{46}L \\
&= \frac{25}{46}L \\
&= 0.543L
\end{aligned}
$$

That's where the balls will meet if they are point particles. Each ball will actually be at the locations $x = 0.54L - R$ and $x = 0.54L + R$.

It is also possible to solve this problem using VPython. We can model the motion of each ball and print the position of the balls when they collide. Here is a sample program.

```
from __future__ import division, print_function
from visual import *

L=1
R=L/100
oofpez=9e9
q1=1e-5
q2=-1e-5
m=1

ball1=sphere(pos=(0,0,0), radius=R, color=color.magenta, make_trail=True)
ball2=sphere(pos=(L,0,0), radius=R, color=color.cyan, make_trail=True)
```

```
I1=2/5*m*R**2
I2=2/3*m*R**2

ball1.v=vector(0,0,0)
ball2.v=vector(0,0,0)

dt=0.0005
t=0


while 1:
    rate(200)
    r=ball1.pos-ball2.pos
    rmag=mag(r)
    runit=r/rmag
    F1=oofpez*q1*q2/rmag**2 * runit

    ball1.v=ball1.v+F1/(m+I1/R**2)*dt
    ball1.pos=ball1.pos+ball1.v*dt

    F2=-F1
    ball2.v=ball2.v+F2/(m+I2/R**2)*dt
    ball2.pos=ball2.pos+ball2.v*dt

    #check to see if centers of balls cross paths
    if(ball2.pos.x<ball1.pos.x):

        #move the balls to their previous positions before the collision
        ball1.pos=ball1.pos-ball1.v*dt
        ball2.pos=ball2.pos-ball2.v*dt

        print(ball1.pos.x)
        break

    t=t+dt
```

The position at the last time step before impact is $0.542L$ which agrees with our previous calculation.

P77:
Solution:

Consider the electron to orbit the nucleus in a circular orbit of radius r with a speed much less than c. Apply the momentum principle to the electron. The only force on the electron is the electric force F by the nucleus.

$$\left| F_{net} \right| = p\frac{v}{r}$$
$$F = \frac{mv^2}{r}$$

In the Bohr model, the angular momentum of the electron is quantized and equal to $L = N\hbar$. Since $L = rp = rmv$ for a circular orbit, then

$$
\begin{aligned}
rmv &= N\hbar \\
v &= \frac{N\hbar}{mr}
\end{aligned}
$$

Substitute v into the momentum principle.

$$
\begin{aligned}
F &= \frac{mv^2}{r} \\
&= \frac{m}{r}\left(\frac{N\hbar}{mr}\right)^2 \\
&= \frac{N^2\hbar^2}{mr^3}
\end{aligned}
$$

The force F is the electric force on the electron by the nucleus. The electron's charge has a magnitude e, and in this case, the nucleus has a charge $2e$.

$$
\begin{aligned}
F &= \frac{1}{4\pi\varepsilon_o}\frac{q_1 q_2}{r^2} \\
F &= \frac{1}{4\pi\varepsilon_o}\frac{2e^2}{r^2}
\end{aligned}
$$

Substitute the electric force into the momentum principle and solve for r.

$$
\begin{aligned}
F &= \frac{N^2\hbar^2}{mr^3} \\
\frac{1}{4\pi\varepsilon_o}\frac{2e^2}{r^2} &= \frac{N^2\hbar^2}{mr^3} \\
r &= \frac{N^2\hbar^2}{2me^2\left(\frac{1}{4\pi\varepsilon_o}\right)}
\end{aligned}
$$

We know from the hydrogen atom that

$$
\frac{\hbar^2}{me^2\left(\frac{1}{4\pi\varepsilon_o}\right)} = 0.529\times10^{-10}\text{ m}
$$

which is called the Bohr radius. Thus, for the singly ionized helium atom

$$
\begin{aligned}
r &= \frac{1}{2}(0.529\times10^{-10}\text{ m})N^2 \\
r &= (0.264\times10^{-10}\text{ m})N^2
\end{aligned}
$$

That fact that the ground state orbital radius is less for singly ionized helium than for hydrogen makes sense. Helium has two protons which exert twice as large a force on the electron than hydrogen. This causes the electron to orbit closer to the nucleus than for hydrogen.

Now let's calculate the energy levels. The energy of the electron and helium nucleus is

$$
\begin{aligned}
E &= K + U_{elec} \\
&= \frac{1}{2}mv^2 + \frac{1}{4\pi\varepsilon_o}\frac{(+2e)(-e)}{r} \\
&= \frac{1}{2}Fr + \frac{1}{4\pi\varepsilon_o}\frac{(+2e)(-e)}{r} \\
&= \frac{1}{2}\frac{1}{4\pi\varepsilon_o}\frac{2e^2}{r} - \frac{1}{4\pi\varepsilon_o}\frac{2e^2}{r} \\
&= -\frac{1}{4\pi\varepsilon_o}\frac{e^2}{r}
\end{aligned}
$$

Substitute the radii of the orbits. Then

$$
\begin{aligned}
E &= -\frac{1}{4\pi\varepsilon_o}\frac{e^2}{r} \\
&= -\frac{1}{4\pi\varepsilon_o}\frac{e^2}{(0.2645 \times 10^{-10}\text{ m})N^2} \\
&= \frac{-8.71 \times 10^{-18}\text{ J}}{N^2}\left(\frac{1\text{ eV}}{1.6 \times 10^{-19}\text{ J}}\right) \\
&= \frac{-54.4\text{ eV}}{N^2}
\end{aligned}
$$

The ground state ($N = 1$) is -54.4 eV. The first excited state ($N = 2$) is -13.6 eV. The energy of a photon emitted in a transition from $N = 2$ to $N = 1$ is $54.4 - 13.6 = 40.8$ eV.

These results differ from hydrogen in that each energy state for the He$^+$ atom is 4 times lower than the equivalent state for hydrogen. The highest energy photon emitted (for a transition from $N = 2$ to $N = 1$) for the He$^+$ atom is 4 times greater than the equivalent transition for hydrogen.

P83:
Solution:

Since it does not precess without the added weight, the gravitational force on the rotor exerts no torque on the rotor (basically, there is another torque that cancels, or balances, this torque). So the only torque on the rotor is due to the weight of mass hanging at the end of the rotor.

(a) To determine which way it precesses, do the following:

(1) Sketch $\vec{L}_i$

(2) Sketch $\Delta\vec{L}$ in the direction of $\vec{\tau}$ due to the hanging mass. Using $\vec{r} \times \vec{F}$ gives $\Delta\vec{L}$ into the page.

(3) Add $\vec{L}_i + \Delta\vec{L}$ to get $\vec{L}_f$. Note that, since $\Delta\vec{L}$ is into the page, then $\vec{L}_f$ has a component into the page.

(4) Determine which way the rotor has precessed. The direction of of rotation of the rotor means that $\vec{L}$ points "down the axis" toward the pivot. If $\vec{L}$ gets a component into the page and also points "down the axis" then the top end must move out of the page. Thus, it precesses clockwise if viewed from above.

(b) The rotational angular momentum of the rotor about its center of mass is

$$\left|\vec{L}_{rot}\right| = I\omega$$

The torque on the rotor due to the mass at its end about the CM of the rotor is

$$\tau_z = -\left|\vec{F}\right| r_\perp$$
$$= -mgL\sin 30°$$

where L is the distance from the CM of the rotor to the top end where m is hanging. Because this is the torque on the system about the center of mass, then the torque produces a change in rotational angular momentum according to the angular momentum principle.

$$\vec{\tau}_{CM} = \frac{d\vec{L}_{rot}}{dt}$$

Because the torque is perpendicular to the rotational angular momentum, the magnitude of $\vec{L}_{rot}$ stays constant; however, the direction of $\vec{L}_{rot}$ about the CM changes due to the torque. Since $\vec{L}_{rot} = I\vec{\omega}$ has a constant magnitude, then $\frac{d\vec{L}_{rot}}{dt} = \left|\vec{L}_{rot}\right| \frac{d\hat{L}}{dt}$.

To find $\frac{d\hat{L}}{dt}$, consider precession of a small angle $\Delta\phi$ in the plane of the circular path of the CM of the gyroscope. This angle is

$$\Delta\phi = \frac{\Delta\vec{L}_{rot}}{L_{rot}\sin(\theta)}$$

where $L_{rot}\sin(\theta)$ is the component of $\vec{L}_{rot}$ in the plane of the circular path of the CM of the gyroscope. Writing $\vec{L}_{rot} = L_{rot}\hat{L}_{rot}$ and dividing both sides by Δt gives

$$\Delta\phi = \frac{\Delta\hat{L}_{rot}}{\sin(\theta)}$$
$$\frac{\Delta\phi}{\Delta t} = \frac{1}{\sin(\theta)}\frac{\Delta\hat{L}_{rot}}{\Delta t}$$

The precessional frequency Ω is $\frac{\Delta\phi}{\Delta t}$. Thus,

$$\Omega = \frac{1}{\sin(\theta)} \frac{\Delta \hat{L}_{rot}}{\Delta t}$$

$$\frac{\Delta \hat{L}_{rot}}{\Delta t} = \Omega \sin(\theta)$$

Now we can apply the angular momentum principle about the CM of the gyroscope.

$$|\vec{\tau}_{CM}| = \left| \frac{d\vec{L}_{rot}}{dt} \right|$$

$$= \left| L_{rot} \frac{d\hat{L}_{rot}}{dt} \right|$$

$$= (I\omega)(\Omega \sin(\theta))$$

$$mgL \sin \theta = (I\omega)(\Omega \sin(\theta))$$

$$\Omega = \frac{mgL}{I\omega}$$

$$\Omega = \frac{(0.2 \text{ kg})(9.8 \frac{\text{N}}{\text{kg}})(0.18 \text{ m})}{(0.06 \text{ kg} \cdot \text{m}^2)(30 \text{ rad/s})}$$

$$= 0.196 \text{ rad/s}$$

The time for 1 revolution is $T = \frac{2\pi}{\Omega} = 32$ s. That's about 0.5 minute, which seems like a slowly precessing gyroscope.

CP87:

Solution:

See the output window after running the program. Here is the program:

```
from __future__ import division , print_function
from visual import *
from visual.graph import *

# setup graph
graph2 = gdisplay(x=430,y=0,width=600,height=400,
                title='Angular displacement vs Time',
                xtitle='Time (s)',
                ytitle='theta (rad)',
                background=color.black)
Tplot = gcurve(color=color.blue)

M = 2
Lrod = 1
R = 0.1
Laxle = 4*R
I = (1/12)*M*Lrod**2 + (1/4)*M*R**2

rod = cylinder(pos=vector(-1,0,0),
            radius=R, color=color.orange,
            axis=vector(Lrod,0,0))
```

```
axle = cylinder(pos=vector(-1+Lrod/2,0,-Laxle/2),
                radius=R/6, color=color.red,
                axis=vector(0,0,4*R))

L = vector(0,0,0) # angular momentum
deltat = 0.0001    # for accuracy in later parts
t = 0
theta = 0
dtheta = 0
p = vector(0,0,0) # linear momentum

axis_of_rotation = vector(0,0,1)
while t < 7:
    rate(10000)
    force = vector(0.1,0,0) # constant force on axle
    # Apply Momentum Principle to the axle and rod
    p = p + force * deltat
    axle.pos = axle.pos + (p/M) * deltat
    rod.pos = rod.pos + (p/M) * deltat
    #torque = vector(0,0,2) # constant torque
    torque = vector(0,0,3*cos(5*t)) # oscillating torque
    # Apply Angular Momentum Principle
    L = L + torque * deltat
    # Update angle and rod position
    omega = L / I
    omega_scalar = dot(omega,norm(axis_of_rotation))
    dtheta = omega_scalar * deltat
    rod.rotate(angle=dtheta, axis=axis_of_rotation,
               origin=axle.pos)
    theta = theta + dtheta
    Tplot.plot(pos=(t,theta))
    t = t + deltat

print ("""Applying a constant force to the axle doesn't
change the rod's oscillatory bahavior.""")
```

12   Chapter 12 Entropy: Limits on the Possible

Q01:

Solution:

The energy distribution has a very low probability of occurring, so we would not expect to observe it in practice.

Q05:

Solution:

According to the Second Law of Thermodynamics, the entropy of the (closed) system of the blocks will increase until it reaches a maximum. Thus, the entropy of the system at any time after the blocks come into contact will be greater than $S_1 + S_2$. Thus, $S > 45$ J/K.

Q09:

Solution:

Air, which is mostly nitrogen, has more internal energy because there are more degrees of freedom (vibrational and rotational) than for helium.

Q13:

Solution:

A and F are the correct choices.

P17:

Solution:

If you flip 10 coins and 5 of them come out to be heads, then there are 10! different arrangements of the coins. But there are 5! permutations of heads among each other and 5! permutations of tails among each other. Therefore, the number of distinct ways to get 5 heads is

$$\Omega = \frac{10!}{5!5!} = \frac{20}{2} = 252$$

For no heads, there is only 1 possible arrangement.

P23:

Solution:

You will have to watch it for 1.7×10^{96} s. This is $(1.7 \times 10^{96} \text{ s}) \left(\frac{1 \text{ year}}{\pi \times 10^7 \text{ s}} \right) = 5.4 \times 10^{88}$ year. This is $\frac{5.4 \times 10^{88} \text{ year}}{10^{10}} \approx 5 \times 10^{78}$ times the age of the Universe.

P27:

Solution:

Since ΔT is small, assume $T \approx T_i$.

For copper $T_i = 50^{\circ}\text{C} + 273\text{K} = 323$ K.

For aluminum $T_i = 45^{\circ}\text{C} + 273\text{K} = 318$ K.

(a)

$$\begin{aligned}
\Delta S_{Al} &= \frac{Q}{T} \\
&\approx \frac{2500 \text{ J}}{318 \text{ K}} \\
&\approx 7.86 \text{ J/K}
\end{aligned}$$

(b)

$$\begin{aligned}
\Delta S_{Cu} &= \frac{Q}{T} \\
&\approx \frac{-2500 \text{ J}}{323 \text{ K}} \\
&\approx -7.74 \text{ J/K}
\end{aligned}$$

(c) The container is insulated. Therefore, there is no interaction with the surroundings, and it is a closed system. Thus, the "Universe" in this case is the system of the aluminum and copper blocks.

$$\begin{aligned}
\Delta S_{sys} &= \Delta S_{Al} + \Delta S_{Cu} \\
&= 7.86 \text{ J/K} - 7.74 \text{ J/K} \\
&= 0.12 \text{ J/K}
\end{aligned}$$

The entropy of the Universe increased as expected according to the Second Law of Thermodynamics.

(d) The change in the energy of the Universe is

$$\begin{aligned}
\Delta E_{sys} &= \Delta E_{Al} + \Delta E_{Cu} \\
&= 2500 \text{ J} + -2500 \text{ J} \\
&= 0
\end{aligned}$$

The energy of the Universe is constant, as expected from the First Law of Thermodynamics.

P33:
Solution:

(a) The energy of one quantum is also the change in energy of the system when it gains one quantum of energy, so $\Delta E = 4 \times 10^{-21}$ J.

For 5 quanta and 18 oscillators, the number of ways to distribute the energy is

$$\begin{aligned}
\Omega &= \frac{(5+18-1)!}{5!17!} \\
&= 26334
\end{aligned}$$

and the entropy is

$$S = k\ln(\Omega)$$
$$= (1.381 \times 10^{-23} \frac{J}{K})\ln(26334)$$
$$= 1.4054 \times 10^{-22} \text{ J/K}$$

For 6 quanta distributed among the same number of oscillators, the number of ways to distribute the energy is

$$\Omega = \frac{(6+18-1)!}{6!17!}$$
$$= 100947$$

and the entropy is

$$S = k\ln(\Omega)$$
$$= (1.381 \times 10^{-23} \frac{J}{K})\ln(100947)$$
$$= 1.591 \times 10^{-22} \text{ J/K}$$

The change in entropy for a change in energy of one quanta is $\Delta S = 1.591 \times 10^{-22} \text{ J/K} - 1.4054 \times 10^{-22} \text{ J/K} = 1.856 \times 10^{-23} \text{ J/K}$.

The temperature is approximately

$$\frac{1}{T} \approx \frac{\Delta S}{\Delta E_{int}}$$
$$T \approx \frac{\Delta E_{int}}{\Delta S}$$
$$\approx \frac{4 \times 10^{-21} \text{ J}}{1.856 \times 10^{-23} \text{ J/K}}$$
$$\approx 215.5 \text{ K}$$

(b) Repeat the above calculations for 8 and 9 quanta of energy. The change in energy (1 quantum) is the same, but we need to calculate the change in entropy.

For 8 quanta, $\Omega = 1,081,575$ and $S = 1.9183 \times 10^{-22} \text{ J/K}$.

For 9 quanta, $\Omega = 3,124,550$ and $S = 2.0648 \times 10^{-22} \text{ J/K}$.

This gives an approximate temperature of $T \approx \frac{4\times10^{-21} \text{ J}}{1.465\times10^{-23} \text{ J/K}} = 273.1 \text{ K}$.

(c) The heat capacity (per atom) is

$$C = \frac{\Delta E_{atom}}{\Delta T}$$

where $\Delta E_{atom} = \frac{\Delta E_{system}}{N_{atoms}}$. The system in this case is the nanoparticle consisting of 6 atoms, and the change in the energy of the nanoparticle is in steps of one quantum of energy, $\Delta E_{nanoparticle} = (8.5-5.5)(4\times10^{-21} \text{ J}) = 1.2\times10^{-20} \text{ J}$. So, the change in the energy of an atom is $(1/6)(1.2 \times 10^{-20} \text{ J}) = 2 \times 10^{-21} \text{ J}$.

The heat capacity in this range of energies is approximately

$$
\begin{aligned}
C &= \frac{\Delta E_{\text{atom}}}{\Delta T} \\
&= \frac{2 \times 10^{-21}\ \text{J}}{(273.1 - 215.5)\ \text{K}} \\
&= 3.47 \times 10^{-23}\ \text{J/K/atom}
\end{aligned}
$$

This is less than the high temperature limit of $3k = 4.2 \times 10^{-23}$ J/K as expected.

P37:

Solution:

Using the density, mass, and Avogadro's number, you can calculate the diameter of a copper atom, assuming a closely packed balls model. Using Young's modulus and the ball-spring model of a solid, you can calculate the spring stiffness.

Calculate the mass of one copper atom.

$$
\begin{aligned}
m_{\text{Cu}} &= \frac{63\ \text{g/mol}}{6.022 \times 10^{23}\ \text{mol}^{-1}} \\
m_{\text{Cu}} &\approx 1.05 \times 10^{-22}\ \text{g}
\end{aligned}
$$

Now, use the density (ρ) and the atom's mass to calculate an approximate interatomic spacing, assuming a cubic atom.

$$
\begin{aligned}
d &\approx \sqrt[3]{\frac{m_{\text{Cu}}}{\rho}} \\
d &\approx \sqrt[3]{\frac{1.05 \times 10^{-22}\ \text{g}}{9\ \text{g/cm}^3}} \approx 2.27 \times 10^{-8}\ \text{cm} \approx 2.27 \times 10^{-10}\ \text{m}
\end{aligned}
$$

Finally, use Young's modulus and interatomic spacing to calculate the interatomic stiffness.

$$
\begin{aligned}
k_{\text{s}} &\approx Y d \\
k_{\text{s}} &\approx \left(1.2 \times 10^{11}\ \text{N/m}^2\right)\left(2.27 \times 10^{-10}\ \text{m}\right) \approx 27\ \text{N/m}
\end{aligned}
$$

In the Einstein model of an atom, the spring stiffness if 4 times this, or 108 N/m.

There is a conceptual way to solve the problem. If kT is comparable to a quantum of energy, then the fact that the energy is quantized is not important, and it is effectively the high temperature limit. In this case,

$$
\begin{aligned}
kT &\approx \hbar\sqrt{k/m} \\
T &\approx \frac{\left(1.055 \times 10^{-34}\ \text{J} \cdot \text{s}\right)\sqrt{(108\ \text{N/m})/(1.05 \times 10^{-25}\ \text{kg})}}{1.381 \times 10^{-23}\ \frac{\text{J}}{\text{K}}} \\
&= 245\ \text{K}
\end{aligned}
$$

The problem can be solved more accurately as well. A general outline of the procedure for solving this problem is given in Section 12.6 of the textbook. It is easiest to write a program and solve it numerically. The program developed for 12.P.72 is a good start and can be modified. The program generally does the following:

1. Begin with zero quanta.

2. Increase the energy by one quantum ($\Delta E = \hbar\sqrt{k/m}$).

3. Calculate the increase in entropy.

4. Calculate the temperature as the slope ($T = \Delta E/\Delta S$).

5. Calculate the specific heat for each change in temperature.

6. Repeat steps (2)-(5).

The reason we're solving it numerically is that it is best to increase the energy only one quantum at a time, and we don't know how many quanta will be needed to reach the high-temperature approximation for C.

A sample program is shown below. The high temperature approximation for C becomes somewhat accurate at around $T=300$ to 400 K.

```
from __future__ import division
from visual import *
from visual.graph import *
from visual.factorial import *


#all data in this program is for copper

w=700 #window width in pixels
maxQuanta = 300 #maximum quanta

#number of atoms in the block
Natoms=35
#number of oscillators in the block
N = 3*Natoms

#constants
kB=1.38e-23 #J/K
hbar = 1.05e-34 #m^2 kg/s
rho=9 #density in g/cm^3
M=63.5 #molar mass in g/mol
NA = 6.022e23 #Avogradro's number in atoms/mol
V=M/rho/NA #volume in cm^3
d=V**(1/3) #diameter in cm
d=d/100 #diameter in m
Y=1.2e11 #Young's modulus
k_s=Y*d #spring stiffness from Y
k = 4*k_s #spring stiffness in Einstein model
m = M/NA/1000 #mass of an atom in kg

#print to check calculations
print d, k, m

#create the graph windows and curves to be graphed
graph1 = gdisplay(width=w, height=w/2, xtitle='T (K)', ytitle='C (J/K/atom)')
curve1 = gcurve(display=graph1.display, color=color.cyan)
data1 = gdots(display=graph1.display, color=color.yellow)
limitCurve1 = gdots(display=graph1.display, color=color.white)
```

```
Nmicrostates=0
qList = []
TList = []

#calculate the temperature of the block of copper
for q in range(0,maxQuanta+1):
    if q>0: #only calculate for q>0
        Si = kB*math.log(omega) #previous entropy for block
    omega=combin((q+N-1),(N-1))
    if q>0:
        Sf = kB*math.log(omega)
        dS = Sf-Si
        T = hbar*sqrt(k/m)/dS
        TList.append(T)
        #Note that there are maxQuanta-1 values of T in the TList

#calculate heat capacity and graph C vs. T
for i in range(1,len(TList)): #note that the list starts with 1 instead of 0
    dT = TList[i] - TList[i-1] #final T - initial T
    C = hbar*sqrt(k/m)/Natoms/dT #C per atom
    curve1.plot(pos=(TList[i],C))
    limitCurve1.plot(pos=(TList[i],3*kB))
```

P43:

Solution:

$$
\begin{aligned}
kT &\approx \left(1.381 \times 10^{-23}\ \frac{\text{J}}{\text{K}}\right)(293\ \text{K}) \\
&\approx 4.04 \times 10^{-21}\ \text{J} \\
&\approx \frac{1}{40}\ \text{eV}
\end{aligned}
$$

P47:

Solution:

The mass of a helium atom is $\left(\frac{0.004\ \text{kg}}{\text{mol}}\right)\left(\frac{1\ \text{mol}}{6 \times 10^{23}\ \text{atoms}}\right) = 6.7 \times 10^{-27}$ kg.

v_{rms} at room temperature is

$$
\begin{aligned}
\bar{v^2} &= \frac{3\,kT}{m} = \frac{3\left(1.381 \times 10^{-23}\ \frac{\text{J}}{\text{K}}\right)(293\ \text{K})}{6.7 \times 10^{-27}\ \text{kg}} = 1.81 \times 10^6\ \text{m}^2/\text{s}^2 \\
v_{rms} &= \sqrt{\bar{v^2}} = 1350\ \text{m/s}
\end{aligned}
$$

The mass of a nitrogen molecule is

$$
m = \left(\frac{0.028\ \text{kg}}{\text{mol}}\right)\left(\frac{1\ \text{mol}}{6 \times 10^{23}\ \text{atoms}}\right) = 4.67 \times 10^{-26}\ \text{kg}
$$

v_{rms} at room temperature is

$$\bar{v^2} = \frac{3\,kT}{m} = \frac{3\left(1.381 \times 10^{-23}\,\frac{J}{K}\right)(293\ K)}{4.67 \times 10^{-26}\ kg} = 2.6 \times 10^5\ m^2/s^2$$

$$v_{rms} = \sqrt{\bar{v^2}} = 510\ m/s$$

P53:
Solution:

(a) The average speed equal to $\int f(v)v\,dv / \int f(v)\,dv$. Since the distribution is not symmetric, the average speed will be slightly greater than the peak. So $v_{avg} \approx 600$ m/s.

(b) v_{rms} is bigger

(c) approximately one tenth

P55:
Solution:

(a) The density of a gas is proportional to $e^{-Mgy/(kT)}$. Assuming constant temperature (which is a poor approximation), the ratio of the density of the atmosphere at height y to the density of the atmosphere at sea level y_0 is

$$\frac{N}{N_0} = \frac{e^{-\frac{Mgy}{kT}}}{e^{-\frac{Mgy_0}{kT}}}$$

$$= e^{-\frac{Mgy}{kT} + \frac{Mgy_0}{kT}}$$

$$= e^{-\frac{Mg}{kT}(y - y_0)}$$

Let's use a single air molecule of diatomic nitrogen N_2 and a temperature of $T = 300$ K, though these are both approximations.

$$\frac{N}{N_0} = e^{-\frac{Mg}{kT}(y - y_0)}$$

$$= e^{-\frac{(4.7 \times 10^{-26}\ kg)(9.8\ N/kg)}{4.14 \times 10^{-21}\ J}(5 \times 10^4\ m)}$$

$$= 0.0038$$

$$= 0.38\%$$

Your answer will depend on what you used for the mass of a molecule of air and what you chose for the air temperature.

(b) The air density at sea level is about $1.23\ kg/m^3$. So the density at an altitude of 50 km would be $0.0038(1.23\ kg/m^3) = 0.0047\ kg/m^3$. Calculate the number of air molecules for this density. Use the molar mass of diatomic nitrogen. In one cubic centimeter of air at this altitude,

$$M = 2\,(14 \text{ g/mol}) \left(\frac{1 \text{ kg}}{1000 \text{ g}} \right) \left(6.022 \times 10^{23} \text{ mol}^{-1} \right) = 4.7 \times 10^{-26} \text{ kg/molecule}$$

$$N = (0.0047 \text{ kg/m}^3) \left(\frac{1 \text{ molecule}}{4.7 \times 10^{-26} \text{ kg}} \right) (1 \text{ cm}^3) \left(\frac{1 \text{ m}^3}{(100)^3 \text{ cm}^3} \right)$$

$$= 1 \times 10^{17} \text{ molecules}$$

(c)

$$\frac{N}{N_0} = e^{-\frac{Mg}{kT}(y - y_0)}$$

$$1 \times 10^{-6} = e^{-\frac{(4.7 \times 10^{-26} \text{ kg})(9.8 \text{ N/kg})}{4.14 \times 10^{-21} \text{ J}}(y - y_0)}$$

$$\ln(1 \times 10^{-6}) = -\frac{(4.7 \times 10^{-26} \text{ kg})(9.8 \text{ N/kg})}{4.14 \times 10^{-21} \text{ J}}(y - y_0)$$

$$(y - y_0) = 1.2 \times 10^5 \text{ m}$$

$$= 120 \text{ km}$$

CP61:
Solution:

(a) The following VPython program can be used to calculate the number of ways to distribute the four quanta among the two atoms and visualize the data as a histogram. It results in the same distribution given in Figure 12.15 of the textbook.

```
from __future__ import division
from visual import *
from visual.graph import *
from visual.factorial import *


w=700 #window width in pixels
quanta=100 #number of quanta
#number of oscillators for each block
N1=300
N2=200

#create the graph window
graph = gdisplay(xmin=0, xmax=quanta, width=w, height=w/2, xtitle='number of quanta
    in block 1', ytitle='total number of microstates')

#create the curve that will be used for the histogram; bars are not used in this
    case
histogram = gvbars(delta=0.5, color=color.red)

Nmicrostates=0

print "quanta in block 1", "quanta in block 2", "total number of microstates"
for q1 in range(0,quanta+1):
    q2=quanta-q1
    omega1=combin((q1+N1-1),(N1-1))
```

```
omega2=combin((q2+N2-1),(N2-1))
Nmicrostates = omega1*omega2
print q1, q2, Nmicrostates
histogram.plot(pos=(q1,Nmicrostates))
```

(b) The following VPython program plots a histogram for the given scenario. The results agree with Fig. 12.21 from the textbook. It's necessary to use the combinatorial function `combin()` in order to calculate the number of microstates. Otherwise, Python will not be able to do the calculation. See the VPython documentation at:

http://vpython.org/contents/docs/visual/factorial.html

for an explanation of the cominatorial and factorial functions.

```
from __future__ import division
from visual import *
from visual.graph import *
from visual.factorial import *


w=700 #window width in pixels
quanta=100 #number of quanta
#number of oscillators for each block
N1=300
N2=200

#create the graph window
graph = gdisplay(xmin=0, xmax=quanta, width=w, height=w/2, xtitle='number of quanta
    in block 1', ytitle='total number of microstates')

#create the curve that will be used for the histogram; bars are not used in this
    case
histogram = gvbars(delta=0.5, color=color.red)

Nmicrostates=0

print "quanta in block 1", "quanta in block 2", "total number of microstates"
for q1 in range(0,quanta+1):
    q2=quanta-q1
    omega1=combin((q1+N1-1),(N1-1))
    omega2=combin((q2+N2-1),(N2-1))
    Nmicrostates = omega1*omega2
    print q1, q2, Nmicrostates
    histogram.plot(pos=(q1,Nmicrostates))
```

The distribution of quanta for which the probability (number of microstates in this case) is half as large as the most probable case (60-40) is the state for which the number of microstates is $(6.866 \times 10^{114})/2 = 3.433 \times 10^{114}$. This is nearest to the probability of finding 66 quanta in the first atom and 34 quanta in the second atom, for which there are 3.76×10^{114} microstates.

(c) Use the program from part (b). Vary $N1$ and $N2$ in the program for different numbers of oscillators. Results for a few choices of $N1$ and $N2$ are shown below.

N_1	N_2	ratio ($N_1 : N_2$)	q_1^*	q_2^*	N^* (# of microstates)
250	250	1:1	50	50	6.73×10^{114}
300	200	3:2	60	40	6.87×10^{114}
400	100	4:1	80	20	8.39×10^{114}
480	20	24:1	97	3	1.74×10^{115}

It's important to try the case where the ratio N_1/N_2 is quite large. You'll notice that the histogram narrows, and its maximum increases as the ratio increases.

13   Chapter 13: Electric Field

Q01:
  Solution:

The relationship is that the force by an electric field on a particle of charge q is $\vec{F}_{\text{by E-field on q}} = q\vec{E}$. The unit of force is newtons (N) and the unit of electric field is newtons per coulomb (N/C).

Q09:
  Solution:

For all of these answers, I will use the convention $\hat{x}$ to the right, $\hat{y}$ toward the top of the page, and $\hat{z}$ outward perpendicular to the page.

(a) $\hat{E} = <0, -1, 0>$. The dipole moment vector $\vec{p}$ points "upward" in the $+y$ direction. Since point A is on the perpendicular bisector of the dipole, the electric field at point A points opposite $\vec{p}$ which is in the $-y$ direction.

(b) $\hat{E} = <0, 1, 0>$. Since point B is along the axis of the dipole, the electric field at point B points in the same direction as the dipole moment, in the $+y$ direction.

(c) Because the electron starts from rest, it will move in the direction of the electric force on the electron which is opposite the electric field. Thus, $\hat{F} = <0, 1, 0>$.

(d) Because it starts from rest, it will move in the direction of the electric force on the proton which is in the same direction as the electric field. Thus, $\hat{F} = <0, 1, 0>$.

(e) The force on the dipole by the electron is opposite the force on the electron by the dipole. Thus the direction of the force on the dipole by the electron is $\hat{F} = <0, -1, 0>$. Because the dipole starts from rest, it will begin moving in the direction $<0, -1, 0>$.

Q15:
  Solution:

The force on the ball by the dipole is directly proportional to the electric field created by the dipole. The electric field by the dipole depends on $1/r^3$, for $r >> s$. If the distance r is doubled, then the electric field changes by a factor $1/2^3 = 1/8$. Therefore the force also changes by $1/8$.

P21:
  Solution:

(a) It is negative since the electric field vectors point toward the charged particle.

(b) The force on the particle at point B is opposite the electric field at point B and thus points down and to the left, radially away from the particle within the dashed circle.

(c)

$$\hat{E} = \frac{\vec{E}}{|\vec{E}|}$$
$$= \frac{<2000, 2000, > \text{ N/C}}{\sqrt{(2000 \text{ N/C})^2 + (2000 \text{ N/C})^2 + (0 \text{ N/C})^2}}$$
$$= <0.707, 0.707, 0>$$

Note that you can determine the answer without doing a calculation because the angle of the vector with respect to the x-axis (or y-axis) is $45°$. As a result the unit vector is $< \cos 45, \cos 45, 0 >$.

(d)

$$
\begin{aligned}
\vec{F} &= q\vec{E} \\
&= (-7 \times 10^{-9} \text{ C})(< 2000, 2000, > \text{ N/C}) \\
&= < -1.4 \times 10^{-5}, -1.4 \times 10^{-5}, 0 > \text{ N}
\end{aligned}
$$

(e) It is opposite the unit vector for the electric field. Thus, $\hat{F} = < -0.707, -0.707, 0 >$. Note that this is down and to the left as predicted in part (b).

P25:

   Solution:

$$
\begin{aligned}
\left| \vec{F}_{net} \right| &= \left| \frac{d\vec{p}}{dt} \right| \\
\left| \vec{F}_{\text{by E-field}} \right| &= m \left| \vec{a} \right| \\
&= (1.673 \times 10^{-27} \text{ kg})(9 \times 10^{11} \text{ m/s}^2) \\
&= 1.50 \times 10^{-15} \text{ N}
\end{aligned}
$$

$$
\begin{aligned}
\left| \vec{F} \right| &= |q| \left| \vec{E} \right| \\
\left| \vec{E} \right| &= \frac{\left| \vec{F} \right|}{|q|} \\
&= \frac{1.50 \times 10^{-15} \text{ N}}{1.602 \times 10^{-19} \text{ C}} \\
&= 9.38 \times 10^{3} \text{ N/C}
\end{aligned}
$$

P33:

   Solution:

$$
\vec{E} = \frac{1}{4\pi\varepsilon_o} \frac{q}{|\vec{r}|^2} \hat{r}
$$

Note that $\vec{r} = \vec{r}_{\text{observation location}} - \vec{r}_{\text{particle}}$.

$$
\begin{aligned}
\vec{r} &= \langle 0.2, 0, 0 \rangle \text{ m} - \langle 0.4, 0, 0 \rangle \text{ m} \\
&= \langle -0.2, 0, 0 \rangle \text{ m} \\
|\vec{r}| &= 0.2 \text{ m} \\
\hat{r} &= \frac{\vec{r}}{|\vec{r}|} \\
&= < -1, 0, 0 >
\end{aligned}
$$

$$\vec{E} = \left(9 \times 10^9 \ \frac{\text{N} \cdot \text{m}^2}{\text{C}^2}\right) \frac{(1 \times 10^{-9} \ \text{C})}{(0.2 \ \text{m})^2} <-1,0,0>$$
$$= \langle -225, 0, 0 \rangle \ \text{N/C}$$

P37:

Solution:

The electric field points in the +x direction and points toward a negative charge. Thus the charged particle must be to the right, in the +x direction, from the point where the electric field is measured.

$$\left|\vec{E}\right| = \frac{1}{4\pi\varepsilon_o} \frac{|q|}{|\vec{r}|^2}$$
$$1 \times 10^3 \ \text{N/C} = \left(9 \times 10^9 \ \frac{\text{N} \cdot \text{m}^2}{\text{C}^2}\right) \frac{1 \times 10^{-6} \ \text{C}}{|\vec{r}|^2}$$
$$|\vec{r}| = 3 \ \text{m}$$

The electric field points toward the charged particle; therefore, the position of the particle is to the right and $\vec{r} = <3, 0, 0>$ m from the location where the electric field is measured.

P43:

Solution:

The electric field points in the +y direction. Since it points toward an electron, then the electron must be in the +y direction from the point where the electric field is measured. The electron must be at a distance given by

$$\left|\vec{E}\right| = \frac{1}{4\pi\varepsilon_o} \frac{|q|}{|\vec{r}|^2}$$
$$160 \ \text{N/C} = \left(9 \times 10^9 \ \frac{\text{N} \cdot \text{m}^2}{\text{C}^2}\right) \frac{\left|-1.602 \times 10^{-19} \ \text{C}\right|}{|\vec{r}|^2}$$
$$|\vec{r}| = 3 \times 10^{-6} \ \text{m}$$

The electric field points toward the electron; therefore, the position of the electron is "upward," and $\vec{r} = <0, 3 \times 10^{-6}, 0>$ m.

P47:

Solution:

(a)

$$\vec{E} = \frac{1}{4\pi\varepsilon_o} \frac{q}{|\vec{r}|^2} \hat{r}$$

Q_1 is at $\vec{r}_{\text{particle}} = \langle 0, 0.03, 0 \rangle$ m, and the observation location is at the location of Q_3 which is $\vec{r}_{\text{observation location}} = \langle 0.04, 0, 0 \rangle$ m.

The position of the observation location relative to the particle is

$$
\begin{aligned}
\vec{r} &= \vec{r}_{\text{observation location}} - \vec{r}_{\text{particle}} \\
&= \langle 0.04, 0, 0 \rangle \text{ m} - \langle 0, 0.03, 0 \rangle \text{ m} \\
&= \langle 0.04, -0.03, 0 \rangle \text{ m}
\end{aligned}
$$

$$
\begin{aligned}
|\vec{r}| &= \left(\sqrt{(0.04)^2 + (0.03)^2 + (0)^2} \right) \text{ m} \\
&= 0.05 \text{ m}
\end{aligned}
$$

$$
\begin{aligned}
\hat{r} &= \frac{\vec{r}}{|\vec{r}|} \\
&= <0.8, -0.6, 0>
\end{aligned}
$$

$$
\begin{aligned}
\vec{E} &= \frac{1}{4\pi\varepsilon_o} \frac{Q_1}{|\vec{r}|^2} \hat{r} \\
&= \left(9 \times 10^9 \ \frac{\text{N} \cdot \text{m}^2}{\text{C}^2} \right) \frac{(-4 \times 10^{-6} \text{ C})}{(0.05 \text{ m})^2} <0.8, -0.6, 0> \\
&= (-1.44 \times 10^7 \text{ N/C}) <0.8, -0.6, 0> \\
&= \langle -1.15 \times 10^7, 8.64 \times 10^6, 0 \rangle \text{ N/C}
\end{aligned}
$$

(b) The electric field due to Q_2 at the location of Q_3 points in the $-y$ direction. Its magnitude is

$$
\begin{aligned}
\left| \vec{E} \right| &= \frac{1}{4\pi\varepsilon_o} \frac{|q|}{|\vec{r}|^2} \\
&= \frac{1}{4\pi\varepsilon_o} \frac{3 \times 10^{-6} \text{ C}}{(0.03 \text{ m})^2} \\
&= 3 \times 10^7 \text{ N/C}
\end{aligned}
$$

So, the E-field vector is $\vec{E} = <0, -3 \times 10^7, 0>$ N/C.

(c) Sum the electric fields due to Q_1 and Q_3.

$$
\begin{aligned}
\vec{E}_{\text{net}} &= \langle -1.15 \times 10^7, 8.64 \times 10^6, 0 \rangle \text{ N/C} + <0, -3 \times 10^7, 0> \text{ N/C} \\
&= \langle -1.15 \times 10^7, -2.13 \times 10^7, 0 \rangle \text{ N/C}
\end{aligned}
$$

(d)

$$
\begin{aligned}
\vec{F}_{\text{net}} &= Q_3 \vec{E} \\
&= (-2 \times 10^{-6} \text{ C}) \left(\langle -1.15 \times 10^7, -2.13 \times 10^7, 0 \rangle \text{ N/C} \right) \\
&= \langle 23.0, 42.6, 0 \rangle \text{ N/C}
\end{aligned}
$$

(e) The electric field due to Q_1 at the location of A points in the $+y$ direction. Its magnitude is

$$
\begin{aligned}
\left|\vec{E}_1\right| &= \frac{1}{4\pi\varepsilon_o}\frac{|q|}{|\vec{r}|^2} \\
&= \frac{1}{4\pi\varepsilon_o}\frac{4\times10^{-6}\ \text{C}}{(0.03\ \text{m})^2} \\
&= 4\times10^7\ \text{N/C}
\end{aligned}
$$

So, the E-field vector is $\vec{E}_1 = <0, 4\times10^7, 0>$ N/C.

(f) The electric field due to Q_2 at the location of A is given by

$$
\begin{aligned}
\vec{r} &= \vec{r}_{\text{observation location}} - \vec{r}_{\text{particle}} \\
&= \langle -0.04, -0.03, 0\rangle\ \text{m}
\end{aligned}
$$

$$
\begin{aligned}
|\vec{r}| &= \left(\sqrt{(0.04)^2 + (0.03)^2 + (0)^2}\right)\ \text{m} \\
&= 0.05\ \text{m}
\end{aligned}
$$

$$
\begin{aligned}
\hat{r} &= \frac{\vec{r}}{|\vec{r}|} \\
&= <-0.8, -0.6, 0>
\end{aligned}
$$

$$
\begin{aligned}
\vec{E}_2 &= \frac{1}{4\pi\varepsilon_o}\frac{Q_2}{|\vec{r}|^2}\hat{r} \\
&= \left(9\times10^9\ \frac{\text{N}\cdot\text{m}^2}{\text{C}^2}\right)\frac{(3\times10^{-6}\ \text{C})}{(0.05\ \text{m})^2}<-0.8, -0.6, 0> \\
&= (1.08\times10^7\ \text{N/C})<-0.8, -0.6, 0> \\
&= \langle -8.64\times10^6, -6.48\times10^6, 0\rangle\ \text{N/C}
\end{aligned}
$$

(g) The electric field due to Q_3 at the location of A points in the $+x$ direction. Its magnitude is

$$
\begin{aligned}
\left|\vec{E}_3\right| &= \frac{1}{4\pi\varepsilon_o}\frac{|Q_3|}{|\vec{r}|^2} \\
&= \frac{1}{4\pi\varepsilon_o}\frac{2\times10^{-6}\ \text{C}}{(0.04\ \text{m})^2} \\
&= 1.125\times10^7\ \text{N/C}
\end{aligned}
$$

So, the E-field vector is $\vec{E}_3 = <1.125\times10^7, 0, 0>$ N/C.

(h) Sum the electric fields due to Q_1, Q_2, and Q_3.

$$
\begin{aligned}
\vec{E}_{net} &= <0, 4 \times 10^7, 0> \text{ N/C} + \langle -8.64 \times 10^6, -6.48 \times 10^6, 0 \rangle \text{ N/C} + <1.125 \times 10^7, 0, 0> \text{ N/C} \\
&= \langle 2.61 \times 10^6, 3.35 \times 10^7, 0 \rangle \text{ N/C}
\end{aligned}
$$

(i)

$$
\begin{aligned}
\vec{F} &= q\vec{E} \\
&= (-3 \times 10^{-9} \text{ C})(\langle 2.61 \times 10^6, 3.35 \times 10^7, 0 \rangle \text{ N/C}) \\
&= \langle -7.83 \times 10^{-3}, -0.101, 0 \rangle \text{ N/C}
\end{aligned}
$$

P53:
Solution:

The point where the field is calculated is along the perpendicular bisector of the dipole. Since $r \gg s$ in this case, calculate the approximate magnitude of E-field at the given point.

$$
\begin{aligned}
\left| \vec{E} \right| &\approx \frac{1}{4\pi\varepsilon_o} \frac{qs}{r^3} \\
&\approx \left(9 \times 10^9 \frac{\text{N} \cdot \text{m}^2}{\text{C}^2} \right) \frac{(1.602 \times 10^{-19} \text{ C})(2 \times 10^{-10} \text{ m})}{(3 \times 10^{-8} \text{ m})^3} \\
&\approx 1.07 \times 10^4 \text{ N/C}
\end{aligned}
$$

P59:
Solution:

Denote the leftmost dipole with subscript 1 and the other dipole with subscript 2.

$$
\begin{aligned}
\vec{E}_{net} &= \vec{E}_1 + \vec{E}_2 \\
&\approx \left\langle 0, \frac{1}{4\pi\varepsilon_o} \frac{|q|s}{d^3}, 0 \right\rangle + \left\langle 0, \frac{1}{4\pi\varepsilon_o} \frac{2|q|s}{d^3}, 0 \right\rangle \\
&\approx \left\langle 0, \frac{1}{4\pi\varepsilon_o} \frac{3|q|s}{d^3}, 0 \right\rangle \\
&\approx \left\langle 0, \left(9 \times 10^9 \frac{\text{N} \cdot \text{m}^2}{\text{C}^2} \right) \frac{3(2 \times 10^{-9} \text{ C})(0.001 \text{ m})}{(0.08 \text{ m})^3}, 0 \right\rangle \\
&\approx \langle 0, 105, 0 \rangle \text{ N/C}
\end{aligned}
$$

P63:
Solution:

First, sketch a picture of the system and label each particle, starting with the particle on the left.

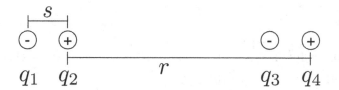

(a) Here are the forces (and net force) acting on particle 1.

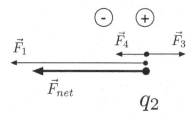

Here are the forces (and net force) acting on particle 2.

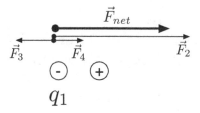

Here are the forces (and net force) acting on particle 3.

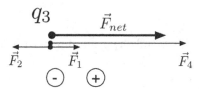

Here are the forces (and net force) acting on particle 4.

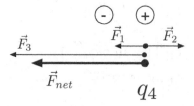

(b) Since we want to calculate the net force by one dipole on the other dipole, let's define a single dipole as the system. In this case, I will choose particles 1 and 2 to be the system. Let's use the naming convention $\vec{F}_{3,1}$ to mean "the force by particle 3 on particle 1." Because we define particles 1 and 2 as the system, the force by particle 1 on particle 2 and the force by particle 2 on particle 1 are internal to the system and are therefore not external forces.

Sketch a free-body diagram showing external forces on the system.

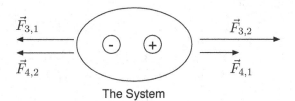

The System

Because of the distances between the particles,

$$\left|\vec{F}_{3,2}\right| > \left|\vec{F}_{4,1}\right|$$

$$\left|\vec{F}_{3,1}\right| = \left|\vec{F}_{4,2}\right|$$

$$\left|\vec{F}_{3,2}\right| > \left|\vec{F}_{3,1}\right|$$

Apply the Principle of Superposition to the system.

$$\vec{F}_{net} = \vec{F}_{3,2} + \vec{F}_{4,1} + \vec{F}_{3,1} + \vec{F}_{4,2}$$

$$= \vec{F}_{3,2} + \vec{F}_{4,1} + 2\vec{F}_{3,1}$$

Define $\hat{r}$ to point to the right along the axis between the dipoles. Then substitute for each electric force on the system.

$$\vec{F}_{net} = \vec{F}_{3,2} + \vec{F}_{4,2} + 2\vec{F}_{3,1}$$

$$= \frac{1}{4\pi\varepsilon_o}\frac{q^2}{(r-s)^2}\hat{r} + \frac{1}{4\pi\varepsilon_o}\frac{q^2}{(r+s)^2}\hat{r} + 2\left(\frac{1}{4\pi\varepsilon_o}\frac{q^2}{r^2}\right)(-\hat{r})$$

$$= \frac{1}{4\pi\varepsilon_o}\frac{q^2}{(r-s)^2}\hat{r} + \frac{1}{4\pi\varepsilon_o}\frac{q^2}{(r+s)^2}\hat{r} - 2\left(\frac{1}{4\pi\varepsilon_o}\frac{q^2}{r^2}\right)\hat{r}$$

$$= \frac{1}{4\pi\varepsilon_o}q^2\left[\frac{1}{(r-s)^2} + \frac{1}{(r+s)^2} - \frac{2}{r^2}\right]\hat{r}$$

Combine terms by finding common denominators. Use algebra to simplify the expression. Start with the first two terms.

$$
\begin{aligned}
\vec{F}_{net} &= \frac{1}{4\pi\varepsilon_o}q^2\left[\frac{(r+s)^2+(r-s)^2}{(r-s)^2(r+s)^2}-\frac{2}{r^2}\right]\hat{r} \\
&= \frac{1}{4\pi\varepsilon_o}q^2\left[\frac{r^2+s^2+2rs+r^2+s^2-2rs}{(r^2-s^2)^2}-\frac{2}{r^2}\right]\hat{r} \\
&= \frac{1}{4\pi\varepsilon_o}q^2\left[\frac{2(r^2+s^2)}{(r^2-s^2)^2}-\frac{2}{r^2}\right]\hat{r} \\
&= 2\frac{1}{4\pi\varepsilon_o}q^2\left[\frac{(r^2+s^2)r^2-(r^2-s^2)}{r^2(r^2-s^2)^2}\right]\hat{r} \\
&= 2\frac{1}{4\pi\varepsilon_o}q^2\left[\frac{r^4+r^2s^2-r^4-s^4+2s^2r^2}{r^2(r^2-s^2)^2}\right]\hat{r} \\
&= 2\frac{1}{4\pi\varepsilon_o}q^2\left[\frac{3s^2r^2-s^4}{r^2(r^2-s^2)^2}\right]\hat{r}
\end{aligned}
$$

Because $r >> s$, then $s^4 \approx 0$ and $r^2 - s^2 \approx r^2$. Then

$$
\begin{aligned}
\vec{F}_{net} &= 2\frac{1}{4\pi\varepsilon_o}q^2\left[\frac{3s^2r^2}{r^6}\right]\hat{r} \\
&= 2\frac{1}{4\pi\varepsilon_o}q^2\left[\frac{3s^2}{r^4}\right]\hat{r} \\
&= \frac{1}{4\pi\varepsilon_o}\left(\frac{6q^2s^2}{r^4}\right)\hat{r}
\end{aligned}
$$

Thus the magnitude of the net force on the dipole by the other dipole is

$$
\left|\vec{F}_{net}\right| = \frac{1}{4\pi\varepsilon_o}\frac{6q^2s^2}{r^4}
$$

CP67:
Solution:

The net electric field at the observation location is $< -33, 541, -112 >$ N/C. A sample program which prints the net electric field at the observation location is shown below.

```
from __future__ import division, print_function
from visual import *

oofpez=9e9

qa=2e-9
particle_a=sphere(pos=(-0.03,0,0.03), radius=0.005, color=color.yellow)

qb=4e-9
particle_b=sphere(pos=(0.03,0,0.03), radius=0.005, color=color.yellow)

qc=-2e-9
```

```
particle_c=sphere(pos=(0,0,-0.011), radius=0.005, color=color.yellow)

E_net = vector(0,0,0) ## initialize E_net

r_obs=vector(0,0.25,0)

r = r_obs - particle_a.pos
rhat = r / mag(r)
E = (oofpez * qa / mag(r)**2) * rhat
E_net = E_net + E

r = r_obs - particle_b.pos
rhat = r / mag(r)
E = (oofpez * qb / mag(r)**2) * rhat
E_net = E_net + E

r = r_obs - particle_c.pos
rhat = r / mag(r)
E = (oofpez * qc / mag(r)**2) * rhat
E_net = E_net + E


sf=0.1/mag(E_net)
ea = arrow(pos=r_obs, axis=sf*E_net, color=color.orange)

print("The net electric field at the observation location is",E_net)
```

Here is the resulting scene.

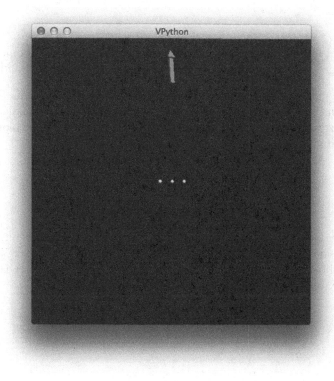

14   Chapter 14: Electric Fields and Matter

Q05:
Solution:

This statement is false. The electric field at the center of the dipole due to the charged particle points to the right, away from the positively charged particle. The electric field at the location of the particle due to the dipole also points to the right because it points in the same direction as the dipole moment of the dipole.

Q13:
Solution:

(a) They swing toward each other because the charged sphere attracts the neutral sphere. After touching, half of the excess electrons are transferred from the charged sphere to the previously neutral sphere. Then, they repel because they have the same charge. Since they are metal spheres, the excess electrons are spread evenly over the entire sphere (as long as they aren't close enough to significantly interact with each other.)

(b) If the block is wide enough to be very near the spheres, then each sphere will be attracted to the neutral block and will touch the block. There will be a larger density of charge on the side of each sphere that is touching the block. If the block is short, then the repulsion between spheres will not be overcome by the much smaller force of attraction between the neutral block and charged spheres.

(c) At point C, a neutral atom will not be polarized because the net electric field due to the spheres is zero. At A, it is polarized with the negative side of the dipole to the right. At B, the induced dipole is like A, but the induced dipole moment is smaller. At point E, the negative side of the induced dipole is shifted to the left. At point D, the induced dipole is oriented like point E except that it has a smaller dipole moment.

Q19:
Solution:

(a) The steel ball and gold foil (basically, the entire electroscope conductor) becomes polarized. The part nearest the rod is negatively charged, and the gold foil leaves become positively charged and repel one another.

(b) Now when close to the glass rod, the foil leaves have less charge. This means that the electroscope is negatively charged. A greater density of electrons is on the top, near the glass rod. A smaller density of electrons on the leaves cause them to still be spread apart, but less than before. This means that the block that contacted the electroscope was negatively charged.

Q25:
Solution:

(a) Sphere B becomes polarized. Since the electron sea in B moves to the right, then the left side of B becomes positively charged and the right side of B becomes negatively charged. When B is polarized, it affects A. The positively charged side of B attracts additional negative charge in A to its right side so that A is no longer uniformly charged. A sketch is shown in the figure below.

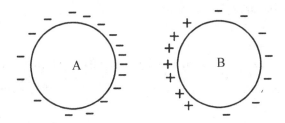

(b) Step 1: The small ball is polarized by both A and B. It is attracted more strongly to A than to B, because A is actually charged negative, whereas B is only polarized, so the ball moves toward A.

Step 2: The ball touches A, and gains negative charge as excess electrons spread out over (A+ball). Now ball is negative, so it is repelled by A, and attracted to B (positive side of B is closer than negative side, so net attraction to B). So the ball moves toward B.

Step 3: The ball touches B, transferring negative charge to B. Now B is slightly negative. The system of (ball+B) is polarized by sphere A, so the ball ends up positive, and is attracted to A, as well as experiencing a net repulsion by B. So the ball moves back toward A.

Step 4: This process continues until A and B have nearly the same amount of charge. At this point, the net force on the ball is no longer large enough to move it toward one sphere, and the process stops.

P29:

Solution:

(2) and (3) are reasonable. Neutrons have no charge, and the energy required to remove a proton from a nucleus is too high to occur with charged tape.

P33:

Solution:

The force by a dipole on a charged particle varies as $1/r^5$. Multiple the distance r by a factor of 3 will change the force by a factor of $1/3^5 = 1/243$. Thus the ratio of the new force to the old force is $F' = F_0/243$.

P37:

Solution:

(a) The force by the polarized carbon atom on the electron has a magnitude

$$F = \left(\frac{1}{4\pi\varepsilon_o}\right)^2 \frac{2\alpha e^2}{r^5}$$

According to the Momentum Principle, $\vec{F}_{net} = m\vec{a}$. Assume that the only force on the electron is that of the carbon atom.

$$\left|\vec{F}\right| = m\left|\vec{a}\right|$$

$$\left|\vec{a}\right| = \frac{\left|\vec{F}\right|}{m}$$

$$= \left(\frac{1}{4\pi\varepsilon_o}\right)^2 \frac{2\alpha e^2}{mr^5}$$

Substitute constants to calculate a.

$$a = \left(9 \times 10^{9} \, \frac{\text{N} \cdot \text{m}^2}{\text{C}^2}\right)^2 \frac{2\left(1.96 \times 10^{-40} \, \text{C} \cdot \text{m}/(\text{N}/\text{C})\right)\left(1.602 \times 10^{-19} \, \text{C}\right)^2}{(9.11 \times 10^{-31} \, \text{kg})(1 \times 10^{-6} \, \text{m})^5}$$
$$\approx 894 \, \text{m/s}^2$$

(b) Since $a \propto 1/r^5$, then doubling the initial distance will change the initial acceleration by a factor $1/2^5 = 1/32$.

P43:
  Solution:

$$u = 5.2 \times 10^{-8} \, \frac{\text{m/s}}{\text{N/C}}$$
$$|\vec{v}| = u\left|\vec{E}\right| = \left(5.2 \times 10^{-8} \, \frac{\text{m/s}}{\text{N/C}}\right)\left(2.4 \times 10^{3} \, \text{N/C}\right)$$
$$|\vec{v}| = 1.25 \times 10^{-4} \, \text{m/s}$$

P47:
  Solution:

b. For a conductor, the charge will be on the surface and the negatively charged electron "sea" shifts away from the electric field, in the direction of the force.

P51:
  Solution:

  (a) 1.

  (b) 3.

  (c) 2.

P57:
  Solution:

  (a)

$$\left|\vec{E}\right| = \frac{1}{4\pi\varepsilon_o} \frac{Q}{|\vec{r}|^2}$$
$$= \left(9 \times 10^{9} \, \frac{\text{N} \cdot \text{m}^2}{\text{C}^2}\right) \frac{\left(3 \times 10^{-9} \, \text{C}\right)}{(0.2 \, \text{m})}$$
$$= 675 \, \text{N/C}$$

(b) Let a be the distance from the origin to the wire's center. The wire becomes polarized such that net electric field at all points inside the wire is zero. Thus, the magnitude of the electric field due to the particle Q must be equal in magnitude to the net electric field due to the polarized charge ($+q$ and $-q$) on the wire. The electric field at the center of a dipole due to the charge ($+q$ and $-q$) of the dipole is:

$$E_{center} = \frac{1}{4\pi\varepsilon_o} \frac{2q}{\left(\frac{s}{2}\right)^2}$$

So at the center of the polarized wire,

$$\frac{1}{4\pi\varepsilon_o} \frac{Q}{a^2} = \frac{1}{4\pi\varepsilon_o} \frac{2q}{\left(\frac{s}{2}\right)^2} = \frac{1}{4\pi\varepsilon_o} \frac{16q}{s^2}$$

Solving for the dipole moment qs gives:

$$qs = Q\frac{s^3}{16a^2}$$

Now we can use the far field approximation to get the electric field due to the polarized wire at the location $< 0.2, 0, 0 >$ m. The distance from the wire to this location is also a. The electric field due to the wire is

$$\left|\vec{E}_{wire}\right| \approx \frac{1}{4\pi\varepsilon_o} \frac{2qs}{a^3}$$

Substitute the equation for qs and solve for the magnitude of the electric field due to the polarized wire.

$$\left|\vec{E}_{wire}\right| \approx \frac{1}{4\pi\varepsilon_o} \frac{Qs^3}{8a^5}$$

$$\approx \left(9 \times 10^9 \; \frac{\text{N} \cdot \text{m}^2}{\text{C}^2}\right) \frac{2\left(3 \times 10^{-9} \; \text{C}\right)\left(1 \times 10^{-3} \; \text{m}\right)^3}{8\left(0.1 \; \text{m}\right)^5}$$

$$\approx 1.82 \times 10^{-2} \; \text{N/C}$$

Because the wire is polarized with $-q$ on the left and $+q$ on the right, the electric field due to the polarized wire at the location $< 0.2, 0, 0 >$ points to the right, away from the wire. Thus, $\vec{E}_{wire} = < 1.82 \times 10^{-2}, 0, 0 >$ N/C.

(c) The electric field due to the particle Q and the wire both point to the right. Thus, the magnitude of the net field will slightly increase due to the presence of the wire.

(d) The wire in no way blocks the particle's electric field due to the particle.

P63:
   Solution:

(a) 2.5 nC, which is half the initial charge of block B. The reason is that the charge distributes itself uniformly across both blocks while they are touching.

(b) 3. Only electrons are mobile (and the conductors do not have positrons).

15 Chapter 15: Electric Field of Distributed Charges

Q01:
Solution:

You can only integrate over the region of space that contains charge, so the limits of integration must be $-\frac{L}{2}$ to $+\frac{L}{2}$.

Q07:
Solution:

The correct order is (3), (2), (4), (1).

Q11:
Solution:

The net field at the ring's center must be zero by symmetry. The field due to any arbitrary chunk will be nulled out by the diametrically opposite chunk's field. The expression derived in the text reduces to zero when the observation location is at the ring's center, just as expected.

Q15:
Solution:

Between the plates, the plates' electric fields are in the same direction and therefore are additive. Outside the plates, the plates' fields are in opposite directions and thus oppose each other. There is a weak dependence on distance that accounts for a small fringe field outside the plates.

P21:
Solution:

(a) charge per length will be $\frac{-Q}{2A}$

(b) $dQ = \frac{-Q}{2A}dx$

(c) $\vec{r} = \vec{r}_{obs} - \vec{r}_{source} = \langle 0, y, 0 \rangle - \langle x, 0, 0 \rangle = \langle -x, y, 0 \rangle$

(d) $|\vec{r}| = \sqrt{x^2 + y^2}$

(e) the integration variable will be x

P25:
Solution:

(a)

$$\vec{F} = -q\vec{E}_- + q\vec{E}_+$$

$$\approx -q\left\langle \frac{1}{4\pi\varepsilon_o} \frac{2(Q/L)}{x - \frac{s}{2}}, 0, 0 \right\rangle + q\left\langle \frac{1}{4\pi\varepsilon_o} \frac{2(Q/L)}{x + \frac{s}{2}}, 0, 0 \right\rangle$$

$$F_x \approx \frac{1}{4\pi\varepsilon_o} \frac{2Qq}{L}\left(-\frac{1}{x - \frac{s}{2}} + \frac{1}{x + \frac{s}{2}} \right)$$

$$F_x \approx \frac{1}{4\pi\varepsilon_o} \frac{2Qq}{L}\left(\frac{-s}{\left(x^2 - \frac{s^2}{4}\right)} \right) \approx -\frac{1}{4\pi\varepsilon_o} \frac{2Q\,|\vec{p}|}{Lx^2}$$

(b)

$$F_x = E_x \frac{|\vec{p}|}{x}$$

$$\approx \frac{\left(1 \times 10^6 \text{ N/C}\right)\left(6 \times 10^{-30} \text{ C} \cdot \text{m}\right)}{\left(1 \times 10^{-2} \text{ m}\right)} \approx 6 \times 10^{-22} \text{ N}$$

$$a_x \approx \frac{F_x}{m} \approx \frac{6 \times 10^{-22} \text{ N}}{\left((0.0018 \text{ kg/mol})/6.022 \times 10^{23} \text{ mol}^{-1}\right)} \approx 2 \times 10^4 \text{ m/s}^2$$

P29:
Solution:

(a) Since the rod is positively charged, each chunk of charge will produce an electric field that points to the right.

(b) L is the rod's length, d is the distance from the right end of the rod to the observation location, x is the distance from the origin to an arbitrary chunk of charge, and $L + d - x$ is the distance from an arbitrary chunk to the observation location. Each chunk produces a field

$$\Delta \vec{E} = \frac{1}{4\pi\varepsilon_o} \frac{\Delta Q}{(L + d - x)^2} \langle 1, 0, 0 \rangle$$

$$= \frac{1}{4\pi\varepsilon_o} \frac{\frac{Q}{L} \Delta x}{(L + d - x)^2} \langle 1, 0, 0 \rangle$$

(c)

$$E = \frac{1}{4\pi\varepsilon_o} \frac{Q}{L} \langle 1, 0, 0 \rangle \int_{x=0}^{x=L} \frac{dx}{(L + d - x)^2}$$

$$= \frac{1}{4\pi\varepsilon_o} \frac{Q}{L} \langle 1, 0, 0 \rangle \int_{u=L+d}^{u=d} \frac{-du}{u^2} = \frac{1}{4\pi\varepsilon_o} \frac{Q}{L} \langle 1, 0, 0 \rangle \frac{1}{d} - \frac{1}{L+d} = \frac{L}{d(L+d)}$$

$$= \frac{1}{4\pi\varepsilon_o} \frac{Q}{d(L + d)}$$

(d) At great distances ($d \gg L$) the field approaches that of a particle as expected.

P33:
Solution:

The ring on the left creates a field at the midpoint between the rings that is to the right. The ring on the right creates a field at the midpoint between the rings that is to the left. These fields are equal in magnitude, so the net electric field at this point equals zero.

P39:
Solution:

(a) The electron sea will shift to the right making the right side slightly negative and the left side slightly positive.

(b) The net electric field inside the foil is zero. $\vec{E}_{disk}$ points to the left and $\vec{E}_{foil}$ points to the right.

(c)

$$
\begin{aligned}
\vec{E}_{net} &= \vec{E}_{disk} + \vec{E}_{foil} \\[4pt]
\left|\vec{E}_{disk}\right| &\approx \frac{|Q|/\pi R^2}{2\varepsilon_o}\left(1 - \frac{d}{R}\right) \\[4pt]
\left|\vec{E}_{foil}\right| &\approx \frac{|q|/\pi r^2}{\varepsilon_o} \\[4pt]
\frac{|q|}{\pi \varepsilon_o r^2} &\approx \frac{|Q|}{2\pi \varepsilon_o R^2}\left(1 - \frac{d}{R}\right) \\[4pt]
|q| &\approx |Q|\frac{r^2}{2R^2}\left(1 - \frac{d}{R}\right) \\[4pt]
&\approx \left|3\times 10^{-5}\ \text{C}\right|\frac{(0.02\ \text{m})^2}{2(1.5\ \text{m})^2}\left(1 - \frac{0.003\ \text{m}}{1.5\ \text{m}}\right) \\[4pt]
&\approx 2.66\times 10^{-9}\ \text{C}
\end{aligned}
$$

P45:
Solution:

(a) As the particle enters the region between the plates, it will deflect downward due to its interaction with the electric field. Upon leaving the plates, its trajectory will not change at all and will remain linear.

(b) At low speeds, $\left|\vec{F}_{net}\right| \approx m\,|\vec{a}|$ or $|\vec{a}| \approx \frac{\left|\vec{F}_{net}\right|}{m} \approx \frac{|q||\vec{E}|}{m} \approx 1.8\times 10^{16}\ \text{m/s}^2$

(c) Treat the plates as a capacitor, and get the magnitude of the charge from $\left|\vec{E}\right| \approx \frac{|Q|/A}{\varepsilon_o}$. The result is $|Q| \approx 3.18\times 10^{-9}$ C. The upper plate must be negatively charged given the direction of $\vec{E}$.

P49:
Solution:

(a) The metal ball will polarize with the right side being more negative than the left side. The molecules in the plastic ball will polarize with their positive ends pointing toward the metal ball.

(b) At the center of the metal sphere, the electric field due to the plastic sphere is

$$
\begin{aligned}
\left|\vec{E}_{plastic}\right| &= 9\times 10^9\ \frac{\text{N}\cdot\text{m}^2}{\text{C}^2}\frac{\left(7\times 10^{-9}\ \text{C}\right)}{\left(9\times 10^{-2}\ \text{m}\right)^2} \approx 7780\ \text{N/C} \\[4pt]
\vec{E}_{plastic} &= \ <-7780, 0, 0>\ \text{N/C}
\end{aligned}
$$

if we neglect the electric field due to the polarized molecules in the plastic.

(c) The net electric field inside the metal ball is zero.

(d) Since the net electric field inside the metal sphere is zero, then $\vec{E}_{charges\ on\ surface} + \vec{E}_{plastic} = 0$. Thus,

$$
\vec{E}_{charges\ on\ surface} = -\vec{E}_{plastic} = <7780, 0, 0>\ \text{N/C}
$$

P51:

   Solution:

At 3 cm from center, $\left|\vec{E}\right|$ is zero because you're inside both shells. At 7 cm from center, the net field is just that of a -25 nC particle, 4.59×10^4 N/C radially inward. At 10 cm from center, the net field is that of a $+39$ nC particle, 3.51×10^4 N/C radially outward.

P57:

   Solution:

   (a) Inside the big sphere, the only contribution to the net field is from the small sphere.

$$E_{A,x} = \frac{1}{4\pi\varepsilon_o} \frac{\left|Q_{small}\right|}{\left|\vec{r}\right|^2}$$

$$\approx \left(9 \times 10^9 \frac{\text{N}\cdot\text{m}^2}{\text{C}^2}\right) \frac{1 \times 10^{-9} \text{ C}}{(0.19 \text{ m})^2} \approx 249 \text{ N/C}$$

$$E_{A,y} = 0$$

   (b) At B, both spheres contribute as though they are particles.

$$\left|\vec{E}_{big}\right| = \left(9 \times 10^9 \frac{\text{N}\cdot\text{m}^2}{\text{C}^2}\right) \frac{4 \times 10^{-9} \text{ C}}{(0.2915 \text{ m})^2} \approx 424 \text{ N/C}$$

$$\left|\vec{E}_{small}\right| = \left(9 \times 10^9 \frac{\text{N}\cdot\text{m}^2}{\text{C}^2}\right) \frac{1 \times 10^{-9} \text{ C}}{(0.15 \text{ m})^2} \approx 400 \text{ N/C}$$

$$\tan\theta \approx \frac{15}{25}$$

$$\sin\theta \approx \frac{15}{29.15}$$

$$\cos\theta \approx \frac{25}{29.15}$$

$$E_{B,x} = \left|\vec{E}_{big}\right|\cos\theta \approx \left(424\frac{25}{29.15}\right) \text{ N/C} \approx 364 \text{ N/C}$$

$$E_{B,y} = \left|\vec{E}_{big}\right|\sin\theta - \left|\vec{E}_{small}\right|$$

$$\approx \left(424\frac{15}{29.15} - 400\right) \text{ N/C} \approx -182 \text{ N/C}$$

   (c)

$$F_{B,x} = -eE_{B,x} \approx \left(-1.602 \times 10^{-19} \text{ C}\right)(364 \text{ N/C}) \approx -5.82 \times 10^{-17} \text{ N}$$

$$F_{B,y} = -eE_{B,y} \approx \left(-1.602 \times 10^{-19} \text{ C}\right)(-182 \text{ N/C}) \approx +2.91 \times 10^{-17} \text{ N}$$

P61:

   Solution:

At the ball's center, the wire creates an electric field $\vec{E}_{wire}$ that polarizes the ball, making it an induced dipole with dipole moment $\vec{p}$. The new dipole ball will create an electric field $\vec{E}_{ball}$ at the wire. The force on the wire due to the dipole (ball) is the charge on the wire multiplied by the dipole's field at the wire. By reciprocity, this is equal to the force on the dipole due to the wire.

$$\left|\vec{E}_{wire}\right| \approx \frac{1}{4\pi\varepsilon_o} \frac{2Q/L}{\left|\vec{d}\right|}$$

$$|\vec{p}| = \alpha\left|\vec{E}\right| wire$$

$$\left|\vec{E}_{ball}\right| \approx \frac{1}{4\pi\varepsilon_o} \frac{2|\vec{p}|}{\left|\vec{d}\right|^3}$$

$$\approx \frac{1}{4\pi\varepsilon_o} \frac{2\alpha\left|\vec{E}_{wire}\right|}{\left|\vec{d}\right|^3}$$

$$\approx \left(\frac{1}{4\pi\varepsilon_o}\right)^2 \frac{4\alpha(Q/L)}{\left|\vec{d}\right|^4}$$

$$\left|\vec{F}_{wire}\right| \approx Q\left|\vec{E}_{ball}\right|$$

$$\approx \left(\frac{1}{4\pi\varepsilon_o}\right)^2 \frac{4\alpha Q^2/L}{\left|\vec{d}\right|^4}$$

$$\approx \left|\vec{F}_{ball}\right|$$

CP63:
Solution:

(1) Approximately 50 points are used.

(2) See program listing.

```
from __future__ import division, print_function
from visual import *

scene.height = 800
oofpez = 9e9
Qtot = 1e-9
L = 1
N = 500
dy = L/N
## create a list of point charges
slices = [] ## an empty list
i = 0
y0 = -L/2 + dy/2 ## center of bottom slice
while i < N:
    a = sphere(pos=vector(0,y0+i*dy,0), radius=dy/2,
            color=color.red, q=Qtot/N)
    slices.append(a) ## add sphere to list
    i = i + 1
```

```
## calculuate E at observation location
r_obs = vector(0.05,0,0) ## to test calculuation
r_obs = vector(0.05,-0.35,0) ## for part (b)
E_net = vector(0,0,0)
i = 0
while i < N:
    rate(100)
    r = r_obs - slices[i].pos
    rhat = r/mag(r)
    E = (oofpez * slices[i].q / mag(r)**2) * rhat
    E_net = E_net + E
    i = i + 1
arrow(pos=r_obs,axis=E_net*0.0002,color=color.orange)
print(E_net)
```

CP67:
Solution:

(a) The electric field away from the ends should be perpendicular to the rod and of uniform magnitude.

(b) See program listing.

```
from __future__ import division, print_function
from visual import *

scene.height = 800
oofpez = 9e9
Qtot = -60e-9
L = 2
N = 50
dz = L/N
## create a list of point charges
slices = [] ## an empty list
i = 0
z0 = -L/2 + dz/2 ## center of bottom slice
while i < N:
    a = sphere(pos=vector(0,0,z0+i*dz), radius=dz/2,
               color=color.red, q=Qtot/N)
    slices.append(a) ## add sphere to list
    i = i + 1
## create a list of arrows
observation = []
dtheta = pi/6
R = 0.30 ## radius of circle
zobs = 0.2
theta = 0
while theta < 2*pi:
    a = arrow(pos=vector(R*cos(theta),R*sin(theta),zobs),
              color=color.orange, axis=vector(0,0,0))
    observation.append(a)
    theta = theta + dtheta
zobs = -0.2
```

```
theta = 0
while theta < 2*pi:
    a = arrow(pos=vector(R*cos(theta),R*sin(theta),zobs),
             color=color.orange,axis=vector(0,0,0))
    observation.append(a)
    theta = theta + dtheta
zobs = 0.4
theta = 0
while theta < 2*pi:
    a = arrow(pos=vector(R*cos(theta),R*sin(theta),zobs),
             color=color.orange,axis=vector(0,0,0))
    observation.append(a)
    theta = theta + dtheta
zobs = -0.4
theta = 0
while theta < 2*pi:
    a = arrow(pos=vector(R*cos(theta),R*sin(theta),zobs),
             color=color.orange,axis=vector(0,0,0))
    observation.append(a)
    theta = theta + dtheta
zobs = 0.6
theta = 0
while theta < 2*pi:
    a = arrow(pos=vector(R*cos(theta),R*sin(theta),zobs),
             color=color.orange,axis=vector(0,0,0))
    observation.append(a)
    theta = theta + dtheta
zobs = -0.6
theta = 0
while theta < 2*pi:
    a = arrow(pos=vector(R*cos(theta),R*sin(theta),zobs),
             color=color.orange,axis=vector(0,0,0))
    observation.append(a)
    theta = theta + dtheta


## calculuate E at observation location
sf = 0.0001 ## arrow scale factor
j = 0
## outer loop
while j < len(observation):
    rate(500)
    earrow = observation[j]
    ## add E of all slices for this obs loc
    i = 0
    E_net = vector(0,0,0)
    ## inner loop
    while i < N:
        r = earrow.pos - slices[i].pos
        rhat = r/mag(r)
        E = (oofpez * slices[i].q / mag(r)**2) * rhat
        E_net = E_net + E
        i = i + 1
```

```
        ## end of inner loop
earrow.axis = sf * E_net
j = j + 1
## end of outer loop
```

16   Chapter 16: Electric Potential

Q03:
Solution:

U_{el} is negative, so q_1 and q_2 are oppositely charged. F and G are possible.

Q11:
Solution:

(a) $\Delta \vec{l} = \vec{r}_A - \vec{r}_B$, so $\Delta \vec{l}$ points in the -y direction, toward the rod.

(b) $\vec{E}$ points toward the rod (in the -y direction). Since $\vec{E}$ and $\Delta \vec{l}$ are in the same direction, $\vec{E} \cdot \Delta \vec{l}$ is positive and $\Delta V = V_A - V_B$ is negative.

Q13:
Solution:

$\vec{dl} = <-dx, 0, 0>$ for path from B to A.

$$\begin{aligned} \Delta V &= -\int \vec{E} \cdot \vec{dl} \\ &= -\int -E_x \, dx \\ &= \int_a^b E_x \, dx \end{aligned}$$

(a) (3)

$$\begin{aligned} \Delta V &= \int_a^b \frac{K}{x^2} \, dx \\ &= \left. \frac{-K}{x} \right|_a^b \\ &= -K\left(\frac{1}{b} - \frac{1}{a}\right) \\ &= K\left(\frac{1}{a} - \frac{1}{b}\right) \end{aligned}$$

(b) (8)

$$\begin{aligned} \Delta V &= \int_a^b \frac{K}{x^3} \, dx \\ &= \frac{-K}{2} \frac{1}{x^2} \Big|_a^b \\ &= \frac{-K}{2} \left(\frac{1}{b^2} - \frac{1}{a^2} \right) \\ &= \frac{K}{2} \left(\frac{1}{b^2} - \frac{1}{a^2} \right) \end{aligned}$$

(c) (6)

$$\begin{aligned} \Delta V &= \int_a^b \frac{K}{x} \, dx \\ &= K \ln x \Big|_a^b \\ &= K(\ln b - \ln a) \\ &= K \ln \frac{b}{a} \end{aligned}$$

(d) (5)

$$\begin{aligned} \Delta V &= \int_a^b Kx \, dx \\ &= \frac{Kx^2}{2} \Big|_a^b \\ &= \frac{K}{2} (b^2 - a^2) \end{aligned}$$

P23:
  Solution:

   (a)

$$\begin{aligned} K &\approx \frac{1}{2} m v^2 \\ &= \frac{1}{2} (9.11 \times 10^{-31} \text{ kg})(6000 \text{ m/s})^2 \\ &= 1.64 \times 10^{-25} \text{ J} \end{aligned}$$

(b)

$$
\begin{aligned}
K &\approx \frac{1}{2}mv^2 \\
&= \frac{1}{2}(1.67 \times 10^{-27} \text{ kg})(6000 \text{ m/s})^2 \\
&= 3.01 \times 10^{-20} \text{ J}
\end{aligned}
$$

Note that the proton has a larger kinetic energy since it has a greater mass.

P27:
  Solution:

$$
\Delta V = -\vec{E} \cdot \Delta \vec{l}
$$

where $\Delta \vec{l}$ is across the plates

$$
\begin{aligned}
\Delta V &= -E_x \Delta x \\
|\Delta V| &= \left|\vec{E}\right| s
\end{aligned}
$$

where s is the plate separation

$$
\begin{aligned}
\left|\vec{E}\right| &= \frac{|\Delta V|}{s} \\
&= \frac{36 \text{ V}}{1 \times 10^{-3} \text{ m}} \\
&= 3.6 \times 10^4 \, \frac{\text{V}}{\text{m}}
\end{aligned}
$$

P33:
  Solution:

$$
\begin{aligned}
\Delta V &= -\vec{E} \cdot d\vec{l} \\
\Delta V &= -E_x \Delta x \\
|\Delta V| &= \left|E_x\right| |\Delta x| \\
\left|E_x\right| &= \frac{2200 \text{ V}}{0.28 \text{ m}} \\
&= 7860 \, \frac{\text{V}}{\text{m}}
\end{aligned}
$$

Note that $E_x = -7860 \, \frac{\text{V}}{\text{m}}$.

P39:
   Solution:

B is at a higher potential since $V_B > V_A$.

$\vec{E}$ points toward lower potential so $\vec{E}$ points to the left in the picture.

$\Delta V = -\int \vec{E} \cdot d\vec{l} = -\vec{E} \cdot \Delta \vec{l}$ for constant electric field. For a path from A to B, $\Delta \vec{l} = <\Delta x, 0, 0>$.

$$\begin{aligned} \Delta V &= -E_x \Delta x \\ E_x &= \frac{-\Delta V}{\Delta x} \\ &= -\frac{1.5 \text{ V}}{0.01 \text{ m}} \\ &= -150 \frac{\text{V}}{\text{m}} \end{aligned}$$

P45:
   Solution:

Consider the proton to be the "source" of the potential.

$$\begin{aligned} \Delta V &= \frac{1}{4\pi\epsilon_0} q \left(\frac{1}{r_f} - \frac{1}{r_i} \right) \\ &= \frac{1}{4\pi\epsilon_0}(1.6 \times 10^{-19} \text{ C}) \left(\frac{1}{6 \times 10^{-10} \text{ m}} - \frac{1}{4 \times 10^{-10} \text{ m}} \right) \\ &= -1.2 \text{ V} \end{aligned}$$

The system is the electron and the potential.

$$\begin{aligned} \Delta U &= q\Delta V \\ &= (-1.6 \times 10^{-19} \text{ C})(-1.2 \text{ V}) \\ &= 1.92 \times 10^{-19} \text{ J} \\ &= 1.2 \text{ eV} \end{aligned}$$

The potential energy increased, which is expected for two oppositely charged particles getting further apart.

P47:
   Solution:

(a) Assume that the thickness of the central plate in negligible. If Δx_1 is the distance from A to the central plate, then

$$\begin{aligned}
\Delta V_1 &= -E_{1,x}\,\Delta x \\
&= -(725\,\frac{V}{m})(0.4\text{ m}) \\
&= -290\text{ V}
\end{aligned}$$

If Δx_2 is from the plate to point B, then

$$\begin{aligned}
\Delta V_2 &= -E_{2,x}\,\Delta x_2 \\
&= -(-425\,\frac{V}{m})(0.2\text{ m}) \\
&= 85\text{ V}
\end{aligned}$$

$$\begin{aligned}
\Delta V &= \Delta V_1 + \Delta V_2 \\
&= -290\text{ V} + 85\text{ V} \\
&= -205\text{ V}
\end{aligned}$$

This is along a path from A to B, so $\vec{V}_B - \vec{V}_A = -205$ V.

(b)

$$\begin{aligned}
\Delta V &= V_A - V_B \\
&= -(V_B - V_A) \\
&= -(-205\text{ V}) \\
&= 205\text{ V}
\end{aligned}$$

(c)

$$\begin{aligned}
\Delta E &= 0 \\
\Delta U + \Delta K &= 0 \\
q\Delta V + \Delta K &= 0 \\
(-e)\Delta V + \Delta K &= 0 \\
\Delta K &= e\Delta V \\
&= (1.6 \times 10^{-19}\text{ C})(-205\text{ V}) \\
&= -3.28 \times 10^{-17}\text{ J}
\end{aligned}$$

(d) The electron must at least reach the center of the plate. After this, the force by the electric field will accelerate it to the right. The minimum kinetic energy at point A in order to just reach the center of the plate (with K=0 at the center of the plate) is

$$
\begin{aligned}
\Delta E &= 0 \\
\Delta U + \Delta K &= 0 \\
q\Delta V + \Delta K &= 0 \\
(-e)\Delta V + \Delta K &= 0 \\
\Delta K &= e\Delta V \\
&= (1.6 \times 10^{-19}\ \text{C})(-290\ \text{V}) \\
K_f - K_i &= -4.64 \times 10^{-17}\ \text{J} \\
0 - K_i &= -4.64 \times 10^{-17}\ \text{J} \\
K_i &= 4.64 \times 10^{-17}\ \text{J} \\
&= 290\ \text{eV}
\end{aligned}
$$

P53:
Solution:

$$
V_A - V_C = (V_B - V_C) + (V_A - V_B)
$$

$\vec{E}$ is in the -x direction, toward the rod. Thus, $V_A - V_B = 0$ since $\vec{E}$ is $\perp$ to $\Delta\vec{l}$.

$$
\begin{aligned}
V_A - V_C &= V_B - V_C \\
&= -E_x\Delta x \\
&= -E_x(x_B - x_C) \\
&= -E_x(b - c) \\
&= -E_x b
\end{aligned}
$$

Substitute the E-field due to the rod for $r \ll L$.

$$
\begin{aligned}
V_A - V_C &\approx -\left(-\frac{1}{4\pi\epsilon_0}\frac{2\frac{Q}{L}}{r}\right)b \\
&\approx \frac{1}{4\pi\epsilon_0}\frac{2\left(\frac{Q}{L}\right)b}{r}
\end{aligned}
$$

P59:
Solution:

(a)

$$
\begin{aligned}
\Delta V &= V_R - V_{\text{origin}} \\
&= \Delta V_{\text{sphere}} + V_{\text{uniform field}}
\end{aligned}
$$

$\Delta V_{\text{sphere}} = 0$ since the charge is uniform on the sphere and $\vec{E}$ due to surface charge is zero inside that sphere.

ΔV due to the uniform field is $\frac{1}{2}(74 \text{ V})$ since the distance from the origin to $r = 5$ m is $\frac{1}{2}$ of the distance to $r = 10$ m and V changes linearly for a uniform E-field. Thus,

$$
\begin{aligned}
\Delta V &= \Delta V_{\text{uniform field}} \\
&= \frac{1}{2}(74 \text{ V}) \\
&= 37 \text{ V}
\end{aligned}
$$

(b)

$$
\begin{aligned}
\Delta V &= \Delta V_{\text{sphere}} + \Delta V_{\text{uniform field}} \\
&= \left(\frac{1}{4\pi\epsilon_0} \frac{-Q}{r_f} - \frac{1}{4\pi\epsilon_0} \frac{-Q}{r_i} \right) + 37 \text{ V} \\
&= \left((9 \times 10^9 \frac{\text{N} \cdot \text{m}^2}{\text{C}^2}) \frac{(-3530 \times 10^{-9} \text{ C})}{10 \text{ m}} - (9 \times 10^9 \frac{\text{N} \cdot \text{m}^2}{\text{C}^2}) \frac{(-3530 \times 10^{-9} \text{ C})}{5 \text{ m}} \right) + 37 \text{ V} \\
&= 3177 \text{ V} + 37 \text{ V} \\
&= 3214 \text{ V}
\end{aligned}
$$

P63:
Solution:

Assume that parts A and B are sufficiently close to the rod and disk that "near field" approximations can be used. Assume no contribution to the E-field from polarized molecules in the glass and plastic.

$$
\begin{aligned}
\Delta V &= V_B - V_A \\
&= \Delta V_{\text{rod}} + \Delta V_{\text{disk}}
\end{aligned}
$$

$$
\begin{aligned}
\Delta V_{\text{rod}} &= -\int \vec{E} \cdot d\vec{l} \\
&= -\int_A^B E_r \, dr \\
&= -\int_A^B \frac{1}{4\pi\epsilon_0} \frac{2\frac{Q}{L}}{r} \, dr \\
&= -\frac{1}{4\pi\epsilon_0} \frac{2Q}{L} \ln r \Big|_A^B \\
&= -\frac{1}{4\pi\epsilon_0} \frac{2Q}{L} (\ln(d-h) - \ln d) \\
&= -\frac{1}{4\pi\epsilon_0} \frac{2Q}{L} \left(\ln \frac{d-h}{d}\right) \\
&= \frac{1}{4\pi\epsilon_0} \frac{2Q}{L} \ln \frac{d}{d-h}
\end{aligned}
$$

$$
\begin{aligned}
\Delta V_{\text{disk}} &= -E_y \Delta y \\
&\approx -\frac{\frac{-Q}{A}}{2\epsilon_0} h \\
&= \frac{\frac{Q}{A}}{2\epsilon_0} h
\end{aligned}
$$

The potential difference due to both the disk and the rod is

$$
\begin{aligned}
\Delta V &= V_B - V_A \\
&\approx \frac{1}{4\pi\epsilon_0} \frac{2Q}{L} \ln\left(\frac{d}{d-h}\right) + \frac{\frac{Q}{A}}{2\epsilon_0} h \\
&\approx \frac{1}{4\pi\epsilon_0} \frac{Q}{R} \ln\left(\frac{d}{d-h}\right) + \frac{\frac{Q}{A}}{2\epsilon_0} h
\end{aligned}
$$

Note that we assumed that $\vec{E}_{\text{disk}} \approx$ constant for points very close to the disk.

P67:

Solution:

   (a) Note that within the shell, V is constant and $\Delta V = 0$. Thus $\Delta V_{A,B} = V_A - V_R$ where V_R is the potential at the surface of the sphere. We have already shown that for a point particle,

$$\Delta V = -\int_{r_i}^{r_f} E_r \, dr$$

$$= -\int_{r_i}^{r_f} \frac{1}{4\pi\epsilon_0} \frac{q}{r^2} \, dr$$

$$= \frac{1}{4\pi\epsilon_0} q \left(\frac{1}{r_f} - \frac{1}{r_i} \right)$$

$$\Delta V_{A,B,C} = \Delta V_{A,B} + \Delta V_{B,C}$$

$$\Delta V_{A,B} = \frac{1}{4\pi\epsilon_0} q \left(\frac{1}{R} - \frac{1}{r_A} \right)$$

$$\Delta V_{B,C} = \frac{1}{4\pi\epsilon_0} q \left(\frac{1}{r_C} - \frac{1}{R} \right)$$

$$\Delta V_{A,B,C} = \frac{1}{4\pi\epsilon_0} q \left(\frac{1}{r_C} - \frac{1}{r_A} \right)$$

Since $r_C = r_A$, then $\Delta V_{A,B,C} = 0$.

(b) $d\vec{l}$ is tangent to the circle and $\vec{E}$ is radial. Thus $\vec{E} \cdot d\vec{l} = 0$ since $\vec{E}$ and $d\vec{l}$ are perpendicular, and $\Delta V = -\int \vec{E} \cdot d\vec{l} = 0$. This agrees with part (a)

P73:
Solution:

$$\Delta V = -\vec{E} \cdot \Delta\vec{l}$$

$$= -(E_x \Delta x + E_y \Delta y + E_z \Delta z)$$

$$= -(E_z \Delta z)$$

$$V_f - V_i = -E_z (z_f - z_i)$$

$$37 \text{ V} - (-36 \text{ V}) = -E_z (3.15 \text{ m} - 3.27 \text{ m})$$

$$73 \text{ V} = -E_z (-0.12 \text{ m})$$

$$E_z = 608 \frac{\text{V}}{\text{m}}$$

It is in the $+z$ direction. This makes sense because $\vec{E}$ points from high potential to low potential and V at $z = 3.27$ m is less than V at $z = 3.15$ m.

P75:
Solution:

$$\Delta V = -\int_{r_i}^{r_f} \vec{E} \cdot d\vec{l}$$

In this case, $d\vec{l} = <0, dy, 0>$ and $E_{y,ring} = -\frac{1}{4\pi\epsilon_0} \frac{Qy}{(a^2+y^2)^{\frac{3}{2}}}$.

$$\begin{aligned}
\Delta V &= -\int_{y_i}^{y_f} -\frac{1}{4\pi\epsilon_0} \frac{Qy}{(a^2+y^2)^{\frac{3}{2}}} \, dy \\
&= \frac{1}{4\pi\epsilon_0} Q \int_{y_i}^{y_f} \frac{y}{(a^2+y^2)^{\frac{3}{2}}} \, dy \\
&= \frac{1}{4\pi\epsilon_0} Q \frac{-1}{(a^2+y^2)^{\frac{1}{2}}} \Bigg|_{y_i}^{y_f} \\
V_f - V_i &= \frac{-1}{4\pi\epsilon_0} Q \left(\frac{1}{(a^2+y_f^2)^{\frac{1}{2}}} - \frac{1}{(a^2+y_i^2)^{\frac{1}{2}}} \right)
\end{aligned}$$

If V_f is at $y_f = \infty$, then $V_f = 0$.

$$\begin{aligned}
0 - V_i &= \frac{-Q}{4\pi\epsilon_0} \left(\cancelto{0}{\frac{1}{\infty}} - \frac{1}{(a^2+h^2)^{\frac{1}{2}}} \right) \\
-V_i &= \frac{1}{4\pi\epsilon_0} \frac{Q}{(a^2+h^2)^{\frac{1}{2}}} \\
V_i &= -\frac{1}{4\pi\epsilon_0} \frac{Q}{(a^2+h^2)^{\frac{1}{2}}}
\end{aligned}$$

Note that V is negative and goes to 0 as $h \to \infty$, as expected.

P83:

Solution:

Assume that the polarization of the plastic does not affect the E-field within regions 1-2 and 3-4 (the vacuum gaps). The E-field in these regions is solely due to the plates. Since the charge on the plates stays the same, then the voltage across the vacuum gaps remains the same. Since the E-field is constant in these regions, the potential is linear and you can use the ratio of the width of the region to the width of the plates. Thus,

$$\begin{aligned}
\Delta V_{12} &= \Delta V_0 \left(\frac{0.5\text{mm}}{2\text{mm}} \right) \\
&= \frac{1}{4}(1000 \text{ V}) \\
&= 250 \text{ V}
\end{aligned}$$

Since region 3-4 has the same width as 1-2, then $\Delta V_{34} = \Delta V_{12} = 250$ V.

Inside the plastic, $\Delta V = \Delta V_{vacuum}/K = \Delta V_{23,plates}/K$. As a result,

$$\Delta V_{23} = \frac{\Delta V_0 \left(\frac{1 \text{ mm}}{2 \text{ mm}}\right)}{K}$$

$$= \frac{1/2(1000 \text{ V})}{5}$$

$$= 100 \text{ V}$$

The total potential difference across the plates is

$$\Delta V_{14} = \Delta V_{12} + \Delta V_{23} + \Delta V_{34}$$

$$= 250 \text{ V} + 100 \text{ V} + 250 \text{ V}$$

$$= 600 \text{ V}$$

As expected, the potential difference decreased as a result of inserting the plastic.

P87:
  Solution:

(a) Note that the E-field due to the dipole in this case is in the $-y$ direction, but the path from A to B is in the $-x$ direction. As a result, $\Delta V = -\int \vec{E} \cdot d\vec{l} = 0$ since $\vec{E}$ and $d\vec{l}$ are perpendicular.

If you use superposition to get V_A, then

$$V_A = V_q + V_{-q}$$

$$= \frac{1}{4\pi\varepsilon_o} \frac{q}{(d^2 + (s/2)^2)^{1/2}} + \frac{1}{4\pi\varepsilon_o} \frac{(-q)}{(d^2 + (s/2)^2)^{1/2}}$$

$$= 0$$

Using a similar calculation, you can easily calculate $V_B = 0$. Then, $\Delta V = V_B - V_A = 0$.

(b)

$$E_x = -\frac{1}{4\pi\varepsilon_o} \frac{2qs}{x^3}$$

$$= -\int_b^a E_x dx$$

$$= -\int_b^a -\frac{1}{4\pi\varepsilon_o} \frac{2qs}{x^3} dx$$

$$= \frac{1}{4\pi\varepsilon_o} 2qs \left(-\frac{1}{2x^2}\right)\Big|_b^a$$

$$= -\frac{1}{4\pi\varepsilon_o} qs \left(\frac{1}{a^2} - \frac{1}{b^2}\right)$$

If you use superposition, then

$$
\begin{aligned}
V_C &= V_{C1} + V_{C2} \\
&= \frac{1}{4\pi\varepsilon_o}\frac{q}{x+s/2} + \frac{1}{4\pi\varepsilon_o}\frac{(-q)}{x+s/2} \\
&= \frac{1}{4\pi\varepsilon_o}q\left(\frac{1}{x+s/2} - \frac{1}{x-s/2}\right) \\
&= \frac{1}{4\pi\varepsilon_o}q\left(\frac{x+s/2-x-s/2}{(x+s/2)(x-s/2)}\right) \\
&= \frac{1}{4\pi\varepsilon_o}q\left(\frac{-s}{(x^2-(s/2)^2)}\right) \\
&\approx -\frac{1}{4\pi\varepsilon_o}\frac{qs}{x^2} \quad \text{since } x \gg s/2 \\
&\approx -\frac{1}{4\pi\varepsilon_o}\frac{qs}{a^2}
\end{aligned}
$$

Likewise, using the same mathematics, $V_D \approx -\frac{1}{4\pi\varepsilon_o}\frac{qs}{b^2}$. Thus,

$$
\begin{aligned}
\Delta V &= V_C - V_D \\
&= -\frac{1}{4\pi\varepsilon_o}\frac{qs}{a^2} - \left(-\frac{1}{4\pi\varepsilon_o}\frac{qs}{b^2}\right) \\
&= -\frac{1}{4\pi\varepsilon_o}qs\left(\frac{1}{a^2} - \frac{1}{b^2}\right)
\end{aligned}
$$

This agrees with the result obtained by integrating $\vec{E}\cdot d\vec{l}$.

(c)

$$
\begin{aligned}
\Delta U &= q\Delta V \\
&= (-e)(V_C - V_D) \\
&= -e\left(-\frac{1}{4\pi\varepsilon_o}qs\left(\frac{1}{a^2} - \frac{1}{b^2}\right)\right) \\
&= \frac{1}{4\pi\varepsilon_o}eqs\left(\frac{1}{a^2} - \frac{1}{b^2}\right)
\end{aligned}
$$

CP91:

Solution:

```
from __future__ import division, print_function
from visual import *
scene.width = scene.height = 700
oofpez = 9e9
L = 0.30  # cylinder length
R = 0.001  # cylinder radius
Q = 4e-8  # ring charge
## Increase the number of particles until the potential difference
## no longer changes to the desired precision.
phinum = 10 # number of angular intervals, particles in each ring
```

```
cnum = 380 # number of rings along x-axis
## source charges
dphi = 2 * pi / phinum
N = cnum * dphi
sources = []
for n in range(1,(cnum+1),1):
    x = -0.5 * L + 0.5 * (2 * n - 1) * L / cnum
    for ph in range(0,phinum,1):
        phi = ph * dphi
        cosphi = cos(phi)
        sinphi = sin(phi)
        a = vector(x,R*cosphi,R*sinphi)
        sources.append(sphere(pos=a,radius=R*dphi,q=Q/N))

## path
A = vector(0,0.05,0)
B = vector(0.15,0.05,0)
path = [A,B]
curve(pos=path, color=color.green) ## draw path
## potentials
VA = 0
VB = 0
i = 0
while i < len(sources):
    rate(500)
    r_A = A - sources[i].pos
    VA = VA + oofpez*sources[i].q/mag(r_A)
    r_B = B - sources[i].pos
    VB = VB + oofpez*sources[i].q/mag(r_B)
    i = i + 1
print('VA=', VA, 'VB=', VB, 'VB-VA=', VB-VA)
```

17   Chapter 17: Magnetic Field

Q05:
Solution:

(a) Conventional current flows south, so beneath the wire, $\vec{B}$ due to the moving electrons is directed east.

(b) The net magnetic field due to Earth and the moving electrons in the wire is the superposition $\vec{B}_{Earth} + \vec{B}_{wire}$, which is northeast.

Q09:
Solution:

Conventional current flows opposite electron current, which in this case is the -z direction.

P15:
Solution:

$\vec{C} = <3, 0, 0>$ and $\vec{D} = <5\cos(30°), 5\sin(50°), 0> = <4.33, 2.5, 0>$.

$$\begin{aligned}
\vec{C} \times \vec{D} &= <(C_y D_z - C_z D_y), (C_z D_x - C_x D_z), (C_x D_y - C_y D_x)> \\
&= <0, 0, 7.5>
\end{aligned}$$

This vector, $\vec{C} \times \vec{D}$, is in the +z direction. Its magnitude is 7.5.

$$\begin{aligned}
\vec{D} \times \vec{C} &= -\vec{C} \times \vec{D} \\
&= <0, 0, -7.5>
\end{aligned}$$

So, $\vec{D} \times \vec{C}$ is in the $-z$ direction and has a magnitude 7.5.

P19:
Solution:

$$\vec{B} = \frac{\mu_o}{4\pi} \frac{q\vec{v} \times \hat{r}}{|\vec{r}|^2}$$

$$\left|\vec{B}\right| = \frac{\mu_o}{4\pi} \frac{|q|\,|\vec{v}|\,|\hat{r}|\sin\theta}{|\vec{r}|^2}$$

Use the right-hand rule in each case to get the direction of $\vec{B}$.

For P_2 and P_5, $\vec{B} = 0$ since $\theta = 0$ and $\theta = 180°$, respectively, and $\sin(0) = \sin(180°) = 0$.

For P_1, $\vec{B}$ is in the -z direction, and

$$\left|\vec{B}\right| = \frac{\mu_o}{4\pi}\frac{(1.6\times 10^{-19}\text{ C})(4\times 10^6\text{ m/s})\sin 30°}{(0.05\text{ m})^2}$$

$$= 1.28\times 10^{-17}\text{ T}$$

$$\vec{B} = \langle 0, 0, -1.28\times 10^{-17}\rangle\text{ T}$$

Also at P_6, $\vec{B} = \langle 0, 0, -1.28\times 10^{-17}\rangle$ T.

At P_3 and P_4, $\vec{B} = \langle 0, 0, 1.28\times 10^{-17}\rangle$ T. Note that $\left|\vec{B}\right|$ is the same at P_1, P_3, P_4, and P_6 due to symmetry.

P23:
Solution:

(a) Direction of $\vec{E}$ is toward the electron. Thus,

$$\hat{E} = <-1, 0, 0>$$

Direction of $\vec{B}$ is given by the right-hand rule. Since the particle is an electron, point your thumb opposite $\vec{v}$, and $\vec{B}$ is tangent to a circle around an axis through your thumb. Thus, $\vec{B}$ is out of the page, in the $+z$ direction. $\hat{B} = <0, 0, 1>$.

(b)

$$\left|\vec{E}\right| = \frac{1}{4\pi\epsilon_o}\frac{|q|}{|\vec{r}|^2}$$

$$= (9\times 10^9\ \frac{\text{N}\cdot\text{m}^2}{\text{C}^2})\frac{(1.6\times 10^{-19}\text{ C})}{(5\times 10^{-10}\text{ m})^2}$$

$$= 5.76\times 10^9\ \frac{\text{N}}{\text{C}}$$

$$\left|\vec{B}\right| = \frac{\mu_o}{4\pi}\frac{|q|\,|\vec{v}|\sin\theta}{|\vec{r}|^2}$$

$$= \frac{(1\times 10^{-7}\ \frac{\text{T}\cdot\text{m}}{\text{A}})(1.6\times 10^{-19}\text{ C})(3\times 10^6\text{ m/s})\sin 60°}{(5\times 10^{-10})^2}$$

$$= 0.66\text{ T}$$

P27:
Solution:

$$\left(\frac{6.02\times 10^{23}\text{ atoms}}{\text{mol}}\right)\left(\frac{1\text{ mol}}{0.059\text{ kg}}\right)\left(\frac{8.8\times 10^3\text{ kg}}{1\text{ m}^3}\right)\left(\frac{1\text{ mobile electron}}{\text{atom}}\right) = 8.98\times 10^{28}\text{ m}^{-3}$$

P33:
  Solution:

(a) The conventional current is in the +x direction, $< 1, 0, 0 >$.

(b) $\Delta l = \frac{1.3\,\text{m}}{8} = 0.1625$ m

(c) $\left| \Delta \vec{l} \right| = 0.1625$ m

(d) $\Delta \vec{l} = (0.1625\text{ m}) < 1,0,0 >= \langle 0.1625, 0, 0 \rangle$ m

(e) $\vec{r}_4 =< \frac{-\Delta l}{2}, 0, 0 >= \langle -0.08125, 0, 0 \rangle$ m

(f)

$$
\begin{aligned}
\vec{r} &= \vec{r}_{obs} - \vec{r}_4 \\
&= \langle 0.081, 0.178, 0 \rangle \text{ m} - \langle -0.08125, 0, 0 \rangle \text{ m} \\
&= \langle 0.16225, 0.178, 0 \rangle \text{ m}
\end{aligned}
$$

(g)

$$
\begin{aligned}
|\vec{r}| &= 0.2409 \text{ m} \\
\hat{r} &= \frac{\vec{r}}{|\vec{r}|} =< 0.674, 0.739, 0 >
\end{aligned}
$$

(h) $\Delta \vec{l} \times \hat{r} = \langle 0, 0, 0.120 \rangle$ m

(i)

$$
\begin{aligned}
\Delta \vec{B} &= \frac{\mu_o}{4\pi} I \frac{\Delta \vec{l} \times \hat{r}}{|\vec{r}|^2} \\
&= \frac{(1 \times 10^{-7}\, \frac{\text{T} \cdot \text{m}}{\text{A}})(6.5 \text{ A})(\langle 0, 0, 0.120 \rangle \text{ m})}{(0.1625 \text{ m})^2} \\
&= \langle 0, 0, 2.95 \times 10^{-6} \rangle \text{ T}
\end{aligned}
$$

Use the right-hand rule to verify the direction. With your thumb pointing to the right and fingers curling around your thumb, your fingers point in the +z direction at point A.

P35:
  Solution:

(a) A sketch of the compass, along with relevant vectors, is shown in the figure below. Current flows south.

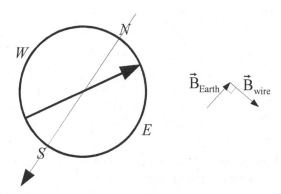

$\vec{B}_{\text{wire}}$ points east, thus the current flows south, according to the right-hand rule.

(b)

$$\tan\theta = \frac{B_{\text{wire}}}{B_{\text{Earth}}}$$

$$B_{\text{wire}} = B_{\text{Earth}}\tan\theta$$

$$\approx (2\times 10^{-5}\text{ T})\tan(10°)$$

$$\approx 3.53\times 10^{-6}\text{ T}$$

$$B_{\text{wire}} = \frac{\mu_o}{4\pi}\frac{2I}{r}$$

$$I = \frac{B_{\text{wire}}\, r}{2\left(\frac{\mu_o}{4\pi}\right)}$$

$$= \frac{(3.53\times 10^{-6}\text{ T})(0.005\text{ m})}{2(1\times 10^{-7}\,\frac{\text{T}\cdot\text{m}}{\text{A}})}$$

$$= 0.088\text{ A}$$

P39:

Solution:

(a) $\vec{B}_{\text{wire}}$ at the location of the compass is west, so I flows N in this part of the wire. Both the top and bottom segments of the wire create a magnetic field at A that is into the page (-z direction).

At the location of the compass, the segment of the wire above the compass dominates. So $\vec{B}_{\text{net}} \approx \vec{B}_{\text{wire}}$ where

$$\vec{B}_{\text{wire}} \approx \frac{\mu_o}{4\pi}\frac{2I}{r}$$

From the compass deflection, $\left|\vec{B}_{\text{wire}}\right| = \left|\vec{B}_{\text{Earth}}\right|\tan\theta = (2\times 10^{-5}\text{ T})\tan(17°) = 6.11\times 10^{-6}\text{ T}$.

Find the current I.

$$I = \frac{\left|\vec{B}_{wire}\right| r}{2\left(\frac{\mu_o}{4\pi}\right)}$$

$$= \frac{(6.11 \times 10^{-6} \text{ T})(0.003 \text{ m})}{2(1 \times 10^{-7} \frac{\text{T} \cdot \text{m}}{\text{A}})}$$

$$= 0.0917 \text{ A}$$

(b) At location A, $\vec{B}$ is the superposition of $\vec{B}$ due to the upper bire and $\vec{B}$ due to the lower wire. Neglect the curved portion of the wire since it is short and far away.

$$\vec{B} = \vec{B}_1 + \vec{B}_2$$

$$\left|\vec{B}_1\right| = \frac{\mu_o}{4\pi} \frac{2I}{r}$$

$$= \frac{(1 \times 10^{-7} \frac{\text{T} \cdot \text{m}}{\text{A}})2(0.0917 \text{ A})}{\left(\frac{0.045 \text{ m}}{2}\right)}$$

$$= 1 \times 10^{-7} \text{ T}$$

$$\vec{B}_1 = \langle 0, 0, -8.15 \times 10^{-7} \rangle \text{ T}$$

Due to symmetry, $\vec{B}_1 = \vec{B}_2$, so

$$\vec{B}_{net} = \vec{B}_1 + \vec{B}_2$$

$$= 2\vec{B}_1$$

$$= \langle 0, 0, -1.63 \times 10^{-6} \rangle \text{ T}$$

(c) We assumed that $r << L$ (very long wire compared to distances where $\vec{B}$ is calculated). We also neglected the curved part of the wire.

P43:

Solution:

The magnetic field at the center due to each current-carrying loop of wire must be in the opposite directions and must be equal in magnitude in order to produce a net magnetic field at the center that is zero. Thus

$$\left| B_{1z} \right| = \left| B_{2z} \right|$$

$$\frac{\mu_0 I_1}{2R_1} = \frac{\mu_0 I_2}{2R_2}$$

$$\frac{I_1}{R_1} = \frac{I_2}{R_2}$$

$$I_2 = I_1 \left(\frac{R_2}{R_1} \right)$$

$$= (6 \text{ A}) \left(\frac{0.02 \text{ m}}{0.08 \text{ m}} \right)$$

$$= \frac{6 \text{ A}}{4}$$

$$= 1.5 \text{ A}$$

P49:
Solution:

(a) Neglect the upper straight line segments because they contribute zero field. Neglect the side segments because they are relatively short and far away and contribute negligible field. Thus, the net magnetic field is a superposition of the magnetic field due to the hemisphere and the lower straight line segment. Application of the right-hand rule shows that the field due to each of these parts of the wire is into the page (-z direction).

(b)

$$\vec{B}_{net} = \vec{B}_{hemisphere} + \vec{B}_{straight \, / \, wire}$$

$$B_{net,z} = B_{z,hemisphere} + B_{z,straight/wire}$$

$$= \frac{\mu_o}{4\pi} \frac{I\pi}{R} + \frac{\mu_o}{4\pi} \frac{2I}{h}$$

where we assume that $h << L$. Thus,

$$\left| \vec{B} \right| = \frac{\mu_o}{4\pi} I \left(\frac{\pi}{R} + \frac{2}{h} \right)$$

P53:
Solution:

$$\left| \vec{B}_{bar} \right| = \left| \vec{B}_E \right| \tan \theta \approx \left(2 \times 10^{-5} \text{ T} \right) \tan 70°$$

$$\approx 5.5 \times 10^{-5} \text{ T}$$

$$\left| \vec{B}_{bar} \right| = \frac{\mu_o}{4\pi} \frac{2 \left| \vec{\mu} \right|}{\left| \vec{r} \right|^3}$$

$$\left| \vec{\mu} \right| = \frac{\left| \vec{B}_{bar} \right| \left| \vec{r} \right|^3}{2 \frac{\mu_o}{4\pi}} \approx \frac{\left(5.5 \times 10^{-5} \text{ T} \right) (0.25 \text{ m})^3}{2 \left(1 \times 10^{-7} \text{ T} \cdot \text{m/A} \right)}$$

$$\approx 4.3 \text{ A} \cdot \text{m}^2$$

P57:

    Solution:

    (a) The field from the coil would deflect the needle in the $+x$ direction, so $\vec{B}_{bar}$ must point in the $-x$ direction. Thus, the bar's S end is closer to the compass.

    (b) Statements 1 and 5 are true.

    (c) The two fields have equal magnitudes and we can set their respective expressions equal to each other to find the coil's magnetic moment.

$$\left|\vec{\mu}_{coil}\right| = \left(\frac{\left|\vec{r}_{coil}\right|}{\left|\vec{r}_{bar}\right|}\right)^3 \left|\vec{\mu}_{bar}\right| \approx \left(\frac{0.097}{0.225}\right)^3 \left(0.54 \text{ A} \cdot \text{m}^2\right)$$

$$\approx 0.043 \text{ A} \cdot \text{m}^2$$

Now use $\left|\vec{\mu}_{coil}\right|$ to solve for the number of turns.

$$N = \frac{\left|\vec{\mu}_{coil}\right|}{AI} = \frac{\left|\vec{\mu}_{coil}\right|}{\pi R^2 I}$$

$$\approx \frac{\left(0.043 \text{ A} \cdot \text{m}^2\right)}{\pi \left(3.5 \times 10^{-2} \text{ m}\right)^2 (0.96 \text{ A})} \approx 12 \text{ turns}$$

CP61:

    Solution:

Here is a sample program. You will notice the value of using a list to store observation locations and arrows. To create the observation locations and the arrows, we use two nested loops where we iterate along x and along theta, respectively. You can change dx and dtheta to adjust proximity and total number of observation locations.

```
from __future__ import division, print_function
from visual import *

scene.x = scene.y = 0
scene.width = 1024
scene.height = 768

mzofp = 1e-7
oofpez = 9e9
L=8e-11

particle = sphere(pos=(-4e-10,0,0), radius=L/10, color=color.red)
particle.v = vector(4e4,0,0)
q = 1.6e-19
dt = 1e-15
Bscale = 1.5e-9

#sketch axes
yaxis=arrow(pos=(0,-2*L,0), axis=4*L*vector(0,1,0), color=color.white, shaftwidth=0.05*
    L, fixedwidth=True)
#sketch axes
```

```
zaxis=arrow(pos=(0,0,-2*L), axis=4*L*vector(0,0,1), color=color.white, shaftwidth=0.05*
    L, fixedwidth=True)

# set up list of observation locations and arrows
R = 8e-11
rlist=[]
xmax=4e-10
dx=1e-10
for x in arange(-xmax,xmax,dx):
        theta=0
        dtheta=45
        while theta<360:
                rlist.append(vector(x, R*cos(theta*pi/180),R*sin(theta*pi/180)))

                theta=theta+dtheta
Barrowlist=[]
for r_obs in rlist:
        Barrowlist.append(arrow(pos=r_obs,
                axis=(0,0,0),color=color.cyan))

# move proton, recalculate all B's at each position
dt = 1e-18
scene.autoscale = 0

while 1:
    rate(2000)
    particle.pos = particle.pos + particle.v * dt
    for Barrow in Barrowlist:
        r = Barrow.pos - particle.pos
        rhat=norm(r)
        B = mzofp*q*cross(particle.v, rhat)/mag(r)**2
        Barrow.axis = B*Bscale
```

Here is a screenshot when the proton passes the origin.

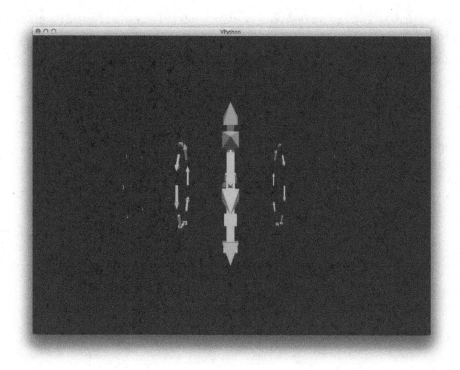

18   Chapter 18: Electric Field and Circuits

Q05:
Solution:

The mobile electrons do not accelerate because the *net force* on the electrons is zero. Though there is an electric force ($\vec{F} = -e\vec{E}$) on the electrons in the direction of their velocity, there is also a force due to drag that is opposite the velocity of the electrons. The sum of these forces is zero, so the acceleration of the electrons is zero. Note that the drag force is due to interactions (i.e. collisions) of mobile electronics with positively charged nuclei and bound electrons.

The dynamics of the situation is similar to a skydiver who jumps out of an airplane. Eventually she reaches a terminal (constant) velocity and her acceleration is zero. The net force on the skydiver is zero because the magnitude of the force of air resistance (drag) on the skydiver is equal to the magnitude of the gravitational force on the skydiver.

Q13:
Solution:

The battery maintains positive surface charge on the positive terminal and negative surface charge on the negative terminal. The electric field within the battery points from the positive terminal toward the negative terminal. Within the wires, the electric field also points away from the positive terminal and toward the negative terminal. However, in a sketch of the circuit, the electric field just inside the battery will be opposite the electric field within the wire just outside the battery.

Q17:
Solution:

The resistor (thin wire) and thick connecting wires are in series. Applying conservation of charge (i.e. the Current Node Rule) to the resistor and wires leads to the conclusion that the electron current through the thin wire must equal the electron current through the thick wire. Use $i = neAuE$ and the fact that n and u are properties of the material in order to compare the quantities given in Table **??**.

resistor	<, =, or >	thick wires
i_R	=	i_w
n_R	=	n_w
A_R	<	A_w
u_R	=	u_w
E_R	>	E_w
v_R	>	v_w

P23:
Solution:

Apply the Current Node Rule (which results from Conservation of Charge flowing into and out of a node).

$i_A = i_B + i_C$

$i_B + i_C = i_D$

$i_A = i_D$

P25:
   Solution:

   (a)

$$\begin{aligned} I_1 + I_4 &= I_2 + I_3 \\ I_2 &= I_1 + I_4 - I_3 \\ &= 7\,\text{A} + 8\,\text{A} - 4\,\text{A} \\ &= 11\,\text{A} \end{aligned}$$

   (b) Yes we did.

   (c)

$$\begin{aligned} I_1 + I_4 &= I_2 + I_3 \\ I_2 &= I_1 + I_4 - I_3 \\ &= 7\,\text{A} + 8\,\text{A} - 20\,\text{A} \\ &= -5\,\text{A} \end{aligned}$$

   (d) No we did not.

P31:
   Solution:

   Note that $i = nAuE$. The electron current is directly proportional to n, A, and u. If you increase A by a factor of 6, for example, then i increases by a factor of 6. Thus, the current in wire B is $i_B = (6)(1.3)(4)i_A = 6.24 \times 10^{19}$ electrons/s.

P35:
   Solution:

   (a) The needle deflects downward.

   (b) The needle deflects 13° upward.

   (c)

$$\begin{aligned} \left| \vec{E}_1 \right| &= \frac{i}{nAu} \\ &\approx \frac{1.5 \times 10^{18} \text{ electrons/s}}{\left(6.3 \times 10^{28} \text{ electrons/m}^3\right)\left(1 \times 10^{-8} \text{ m}^2\right)\left(1.2 \times 10^{-4}\,\frac{\text{m/s}}{\text{N/C}}\right)} \\ &\approx 19.8 \text{ V/m} \end{aligned}$$

   $\vec{E}_1$ points to the left. In bulb 3, the electric field is to the right.

P43:
   Solution:

   (a) (2)

(b)

$$V_G - V_B + V_E - V_G + V_D - V_E + V_B - V_D = 0$$
$$\text{emf}_1 + V_E - V_G + \text{emf}_2 + V_B - V_D = 0$$
$$1.3\ \text{V} + V_E - V_G + 1.3\ \text{V} + V_B - V_D = 0$$
$$2.6\ \text{V} + V_E - V_G + V_B - V_D = 0$$

(c) Since the wires are identical, $V_E - V_G = V_B - V_D$, so

$$2.6\ \text{V} + 2(V_B - V_D) = 0$$
$$V_B - V_D = \frac{2.6\ \text{V}}{2}$$
$$= 1.3\ \text{V}$$

Since E is uniform in the wire, $\Delta V = EL$.

$$E = \frac{V_B - V_D}{L}$$
$$= \frac{1.3\ \text{V}}{0.26\ \text{m}}$$
$$= 5\ \frac{\text{V}}{\text{m}}$$

So, $E_B = 5\ \frac{\text{V}}{\text{m}}$. E at all points in the wire is $5\ \frac{\text{V}}{\text{m}}$.

(d) i is the same everywhere since charge is conserved. Thus,

$$i = nAuE$$
$$= (7 \times 10^{28}\ \text{m}^{-3})(\pi)\left(\frac{7 \times 10^{-4}\ \text{m}}{2}\right)^2\left(5 \times 10^{-5}\ \frac{\text{m/s}}{\text{V/m}}\right)\left(5\ \frac{\text{V}}{\text{m}}\right)$$
$$= 6.73 \times 10^{18}\ \text{s}^{-1}$$

Since $\Delta V = EL$ and ΔV is the same, then E is the same. $E_B = 5\ \frac{\text{V}}{\text{m}}$.

(e) (C)

P47:
Solution:

(a) Since $i_{\text{thin}} = \frac{1}{2}i_{\text{thick}}$ when each bulb is individually connected to the batteries, then $A_{\text{thin}} = \frac{1}{2}A_{\text{thick}}$.

Apply the node rule to the bulbs. Note that experiment showed that the oblong bulb has the thinner filament. Thus,

$$
\begin{aligned}
i_{thin} &= i_{thick} + i_{thick} \\
i_{thin} &= 2i_{thick} \\
nA_{thin}\, uE_{thin} &= 2nA_{thick}\, uE_{thick} \\
A_{thin}\, E_{thin} &= 2A_{thick}\, E_{thick} \\
\tfrac{1}{2}A_{thick}\, E_{thin} &= A_{thick}\, E_{thick} \\
E_{thin} &= 4E_{thick}
\end{aligned}
$$

Apply the loop rule to the circuit.

$$
\begin{aligned}
2\text{emf} - \Delta V_{thin} - \Delta V_{thick} &= 0 \\
2\text{emf} - E_{thin} L_{thin} - E_{thick} L_{thick} &= 0
\end{aligned}
$$

Substitute the result of the node rule.

$$
2\text{emf} - 4E_{thick} L_{thin} - E_{thick} L_{thick} = 0
$$

Assume $L_{thin} = L_{thick}$ (same length filaments), so

$$
\begin{aligned}
2\text{emf} - 4E_{thick} - E_{thick} L &= 0 \\
2\text{emf} - 5E_{thick} &= 0 \\
E_{thick} &= \frac{\tfrac{2}{5}\text{emf}}{L} \\
E_{thin} &= 4E_{thick} = \frac{\tfrac{8}{5}\text{emf}}{L} \\
i_{thin} &= nA_{thin}\, u\left(\frac{\tfrac{8}{5}\text{emf}}{L}\right)
\end{aligned}
$$

When the thin bulb was connected by itself to the batteries, $E_{thin} = \frac{2\text{emf}}{L}$ and $i_{thin} = 1.5 \times 10^{18}\ \text{s}^{-1}$. Calculate $nA_{thin}\, u\frac{\text{emf}}{L}$ with this information.

$$
\begin{aligned}
i_{thin} &= nA_{thin}\, uE_{thin} \\
1.5 \times 10^{18}\ \text{s}^{-1} &= nA_{thin}\, u\frac{2\text{emf}}{L} \\
\frac{2A_{thin}\, u(\text{emf})}{L} &= 0.75 \times 10^{18}\ \text{s}^{-1}
\end{aligned}
$$

Substitute this constant to get i_{thin} for the given circuit.

$$i_{thin} = nA_{thin} u \frac{(\text{emf})}{L} \left(\frac{8}{5} \right)$$

$$= (0.75 \times 10^{18} \text{ s}^{-1}) \left(\frac{8}{5} \right)$$

$$= 1.2 \times 10^{18} \text{ s}^{-1}$$

(b) The biggest approximations we made were (1) to neglect the connecting wires, (2) that the filaments were the same length, and (3) tha battery has no internal resistance (i.e. it is an ideal battery).

(c) The surface charge has a high positive density at the positive terminal of the battery and high negative density at the negative terminal of the battery. Along the connecting wires, the gradient is slight (i.e. it doesn't change very much). The gradient is large along the bulb filaments. The gradient is longest for the oblong bulb (thin filament) since $E_{thin} > E_{thick}$. The gradient along the round bulb is large but not as large as the oblong bulb, since $E_{thick} < E_{thin}$.

P53:
Solution:

The round bulb has a thicker filament. As a result, when it is connected by itself to the battery, the current will be larger and the compass next to the battery will deflect more. Thus, it is the thick-filament bulb that deflects the compass more. For clarity, let's name the oblong bulb 1 and the round bulb 2. Write the loop equation for a single bulb circuit for bulb 1.

$$1: \quad 2\text{emf} - E_1 L = 0$$

$$E_1 = \frac{2\text{emf}}{L}$$

The current through bulb 1 is $i = nAuE_1 = nAu\frac{2\text{emf}}{L}$ when connected by itself to the batteries.

Now apply the loop equation for the two-bulb circuit using a loop consisting of the batteries and bulb 1.

$$2\text{emf} - E_1 L = 0$$

Note that this is exactly this same as when it was the only bulb in the circuit. Thus, the current through the bulb in the two-bulb circuit is the same when bulb 2 is unscrewed. As a result, the compass next to bulb 1 (the oblong bulb) will deflect the same amount, 4°.

The exact same reasoning can be applied to bulb 2, so the compass next to bulb 2 will deflect the same amount as when bulb 1 is unscrewed, which is 15°.

In the two-bulb circuit, applying the current node rule shows that the current through the battery and connecting wires is equal to the sum of the currents through the bulbs, $i = i_1 + i_2$, in according with Conservation of Charge. If the current i_1 creates a magnetic field $\vec{B}_1$ and if the current i_2 creates a magnetic field $\vec{B}_2$, then the total current i will create a magnetic field that is $\vec{B} = \vec{B}_1 + \vec{B}_2$ (since $B \propto i$ for a wire). Note that both of these fields are in the same direction, east, so we can simply add their magnitudes.

The compass next to the battery will deflect an amount θ given by

$$\tan\theta = \frac{B_{wire}}{B_{earth}}$$

$$= \frac{B_1 + B_2}{B_{earth}}$$

$$= \frac{B_{earth}\tan(4°) + B_{earth}\tan(15°)}{B_{earth}}$$

$$= \tan(4°) + \tan(15°)$$

$$= 0.3379$$

$$\theta = \tan^{-1}(0.3379)$$

$$\theta = 18.7°$$

Note that we cannot simply add the angles. The currents add and the magnetic fields (created by those currents) add; however, the angles do not add. But in this case, since $\tan\theta \approx \theta$ for small angles, then $18.7° \approx 4° + 15°$. It is worth noting that for larger currents, simply adding the angles would not give a correct answer. You must add the tangents of the angles.

A sketch of the compasses is shown in the figure below. The deflection angles in the sketch are exaggerated for greater clarity. Note that conventional current flows from the + terminal of one battery to the – terminal of the second battery. As a result, the compass near the + terminal deflects NW since $\vec{B}_{wire}$ is west, and the compass near the – terminal deflects NE since $\vec{B}_{wire}$ is east.

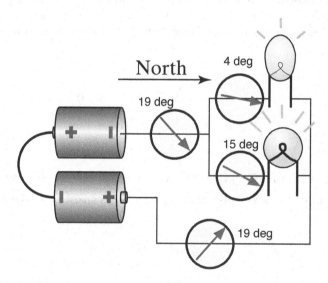

19   Chapter 19: Circuit Elements

Q05:
Solution:

A metal slab is inserted between the plates of a capacitor (such that it doesn't touch the plates). As charge builds up on the plates, the metal slab will become polarized, creating two regions of charge separation across the two gaps between the slab and the plates. The new capacitor is like having two capacitors connected in series, with each one having a gap that is much less than the original capacitor. The fringe field outside of one plate of the capacitor is mostly due to the charge on that plate and the charge on the side of the metal slab nearest that plate.

Suppose that within a fraction of time, an amount of charge ΔQ is added to the plate. Then, the fringe field increases by an amount

$$\Delta E_{fringe} \approx \frac{\Delta Q/A}{\varepsilon}\left(\frac{s}{2R}\right)$$

where s is the distance between the plate and the metal slab. Since this distance is smaller than the plate separation in the original capacitor, then ΔE_{fringe} is less, ΔE_{net} is less, and ΔI is less. As a result, the current is more nearly constant, compared to the original capacitor.

Q11:
Solution:

(a) C. The filament has the greatest effective cross sectional area.

(b) B. The filament has the greatest effective length.

(c) Suggestions for an experimental procedure are given.

Q19:
Solution:

(a) While discharging, conventional current in the wire flows away from the positively charged plate. In the wire beneath the left compass, conventional current runs south, creating a magnetic field above the wire that is west and causes the compass needle to point NW. Conventional current flows north in the wire beneath the right compass. So above the wire at the location of the compass, the current in the wire creates a magnetic field that is east. This causes the right compass to point NE.

When the bulb is connected directly to two batteries, the compasses deflect 15°. Applying the loop rule to the bulb and batteries gives $2\text{emf} = \Delta V_R = IR$. Thus, $I = \frac{2\text{emf}}{R}$.

When the bulb and capacitor are connected directly to two batteries, applying the loop rule to the capacitor and batteries gives $2\text{emf} = \Delta V_C$.

So, when the bulb and capacitor are connected directly together (with no batteries), applying the loop rules at $t = 0$ gives

$$\begin{aligned}\Delta V_C &= \Delta V_R \\ 2\text{emf} &= IR\end{aligned}$$

Thus, at $t = 0$ when first connected, the current through the bulb will be the same as when the bulb was connected directly to the batteries. This will cause the compasses to deflect 15°. As time elapses during the next minute, ΔV_C decreases exponentially, so ΔV_R decreases exponentially. By Ohm's law, I must also decrease exponentially. As a result, the angle of deflection on the compasses will decrease from 15°.

(b) A sketch of electric field vectors is shown in the figure below.

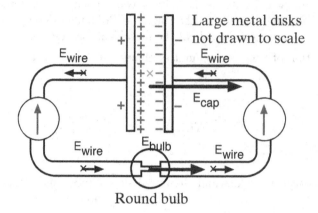

Round bulb

The electric field within the capacitor and the electric field within the bulb filament are much larger than the electric field within the wire.

(c) There is a gradient of positive surface charge from high density at the left (positive) plate to low positive density at the left end of the filament.

There is a gradient of negative surface charge from high negative density at the right (negative) plate to low negative density at the right end of the filament.

The charge density along the filament changes from positive on the left to negative on the right. Note that the gradient (i.e. change) in charge density along the filament is greater than along the wires. This what makes the electric field within the filament larger than in the wires.

The surface charge is depicted in the figure below.

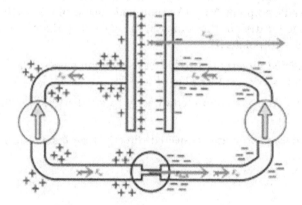

(d) Conservation of charge (the current node rule) applied to the node between the wire and bulb leads to:

$$
\begin{aligned}
I_{bulb} &= I_{wire} \\
neA_{bulb}u_{bulb}E_{bulb} &= neA_{wire}u_{wire}E_{wire} \\
A_{bulb}u_{bulb}E_{bulb} &= A_{wire}u_{wire}E_{wire} \\
d^2_{bulb}u_{bulb}E_{bulb} &= d^2_{wire}u_{wire}E_{wire} \\
\frac{E_{bulb}}{E_{wire}} &= \frac{d^2_{wire}u_{wire}}{d^2_{bulb}u_{bulb}}.
\end{aligned}
$$

Conservation of energy around the circuit (the loop rule) leads to:

$$
\Delta V_C = \Delta V_{wire} + \Delta V_{bulb} + \Delta V_{wire}
$$

At only 0.01 s after connecting the wires, the voltage of the capacitor may not have changed very much from its value at t=0 (depending on its capacitance and the resistance of the bulb). Let's assume that $\Delta V_C \approx 2$emf (although it can be determined more accurately to be $\Delta V_C = 2\text{emf}(1 - e^{t/(RC)})$).

$$
\begin{aligned}
2\text{emf} &\approx 2\Delta V_{wire} + \Delta V_{bulb} \\
2\text{emf} &\approx 2E_{wire}L_{wire} + E_{bulb}L_{bulb}
\end{aligned}
$$

The two valid equations given above that involve the electric fields within the bulb and wire are a result of application of Conservation of charge (node rule) and Conservation of energy (loop rule).

(e) In part(a), we said that the current would decrease and the compass deflections would decrease, starting at 15° and getting smaller as time elapses. The reason is that positive charge (using a conventional current model) leaves the positive plate, causing Q to decrease and the potential difference across the capacitor to decrease. At the same time, positive charge is added to the negative plate causing its charge $-Q$ to also decrease in magnitude. This decreases the surface charge all along the wire. As the surface charge gradient decreases, so does the electric field within the wire and bulb. As the electric field within the wire and bulb decreases, the current decreases.

You can see this mathematically from the loop rule. At every instant t,

$$
\begin{aligned}
\Delta V_C &= 2\Delta V_{wire} + \Delta V_{bulb} \\
&= 2E_{wire}L_{wire} + E_{bulb}L_{bulb}
\end{aligned}
$$

As ΔV_C decreases, E_{wire} and E_{bulb} decrease as well. One might try to argue that only one of the electric fields would have to decrease; however, we know they are related by $\frac{E_{bulb}}{E_{wire}} = \frac{d^2_{wire}u_{wire}}{d^2_{bulb}u_{bulb}}$, so as one of them decreases, the other one must decrease as well.

You've seen in this solution both a qualitative explanation and a quantitative explanation.

Q23:

Solution:

The 60 W bulb has greater resistance. ΔV is the same for each bulb. Substitute $I = \Delta V/R$ (from Ohm's law) into

$P = I\Delta V$, then $P = \frac{\Delta V^2}{R}$. Since $P \propto 1/R$, a larger resistance results in less power. Thus, the less power bulb has a greater resistance.

Q27:
Solution:

An ammeter should have a LOW resistance. The reason is that an ammeter is connected in series with a resistor (or other device) so that the current through the ammeter is the same as the current through the resistor. In order to not affect the current that it is trying to measure, it should have negligible resistance.

A voltmeter should have a HIGH resistance. The reason is that a voltmeter is connected in parallel with a resistor (or other device) so that the voltage (i.e. potential difference) across the voltmeter is the same as the voltage across the resistor. In order to not affect the current through the resistor, it should have a HIGH resistance so that very little current flows through the voltmeter.

P29:
Solution:

(d); Since $Q = Q_0 e^{-\frac{t}{RC}}$ then $|I| = \left|\frac{dQ}{dt}\right| = \frac{Q_0}{RC} e^{-\frac{t}{RC}}$.

P35:
Solution:

1, 3, and 4 are true.

P39:
Solution:

Capacitance is

$$
\begin{aligned}
C &= \frac{K\varepsilon_0 A}{s} \\
&= \frac{(2.9)(8.85 \times 10^{-12} \text{ F/m})(0.59 \text{ m})(0.33 \text{ m})}{0.27 \times 10^{-3} \text{ m}} \\
&= 1.85 \times 10^{-8} \text{ F}
\end{aligned}
$$

P47:
Solution:

Apply the loop rule to the circuit.

$$
\begin{aligned}
2\text{emf} - \Delta V_{\text{R}} &= 0 \\
2\text{emf} &= IR \\
R &= \frac{2\text{emf}}{I} \\
&= \frac{2(1.5 \text{ V})}{0.075 \text{ A}} \\
&= 40 \text{ }\Omega
\end{aligned}
$$

P53:

   Solution:

   A light bulb filament is non-ohmic. Its resistance increases with temperature. A greater current causes the filament to get hotter, which causes resistance to increase.

   When 1 bulb was connected to the battery, it had a greater resistance than when two bulbs in series were connected to the battery.

P59:

   Solution:

   (a) The electric field E_1 is small and uniform throughout the thick wires. From the current node rule applied to the thick and thin wires, we know that the electric field E_2 is much larger and uniform throughout the thin wires, since $neA_1uE_1 = neA_2uE_2$. A sketch is shown in the figure below.

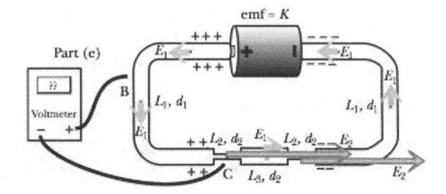

   (b) Roughly, we expect a small gradient along the thick wires (small E_1) and a large gradient along the thin wires (large E_2). By symmetry the central thick wire has very little charge. The surface charge is sketched in the figure below.

   (c) Applying the loop rule (conservation of energy) to the circuit gives:

$$K - \Delta V_1 - \Delta V_2 - \Delta V_3 - \Delta V_1 - \Delta V_2 = 0$$
$$K - 2E_1 L_1 - 2E_2 L_2 - E_1 L_3 = 0$$
$$K = E_1(2L_1 + L_3) + 2E_2 L_2$$

   Using the above equation with $A_1 E_1 = A_2 E_2$ from the current node rule (conservation of charge), solve for the electron current $i = nAuE$. Begin by substituting for E_1.

$$K = E_2\left(\frac{A_2}{A_1}(2L_1 + L_3) + 2L_2\right)$$
$$E_2 = \frac{K}{\left(\frac{A_2}{A_1}(2L_1 + L_3) + 2L_2\right)}$$

   The current through the thin wire i_2 is

$$i_2 = nA_2uE_2$$
$$= \frac{nA_2uK}{\left(\frac{A_2}{A_1}(2L_1 + L_3) + 2L_2\right)}$$

Substitute for the area, $A_2 = (1/4)\pi d_2^2$. Also note that $i_1 = i_2$, so solving for i_2 is the equivalent of solving or i_1. The number of electrons leaving the negative terminal of the battery is

$$i = \frac{n(1/4)\pi d_2^2 uK}{\left(\frac{d_2^2}{d_1^2}(2L_1 + L_3) + 2L_2\right)}$$

By multiplying by $\frac{A_1/A_2}{A_1/A_2}$, you can also write the current as

$$i = \frac{n(1/4)\pi d_1^2 uK}{\left((2L_1 + L_3) + 2\frac{d_1^2}{d_2^2}L_2\right)}$$

(d) If $d_2 << d_1$, then $\frac{d_2^2}{d_1^2} \approx zero$, so the first expression for i gives

$$i \approx \frac{n(1/4)\pi d_2^2 uK}{2L_2}$$

Essentially this is the same result as if we had ignored ΔV for each think wire.

(e) The potential difference read by the voltmeter is $\Delta V = E_2(\frac{L_2}{2})$. The electric field E_2 is approximately $K/(2L_2)$ since the thin wires are in series and we are neglecting the potential drop across the thick wires. Thus, the voltmeter will read $\Delta V = K/(2L_2)(\frac{L_2}{2}) = (1/4)K$. The sign is positive because the positive lead was connected to the high-potential side of the wire.

P65:
   Solution:

$$W = \int_0^Q \frac{q}{C}dq$$
$$= \frac{1}{C}\int_0^Q q\,dq$$
$$= \frac{1}{C}\frac{q^2}{2}\Big|_0^Q$$
$$= \frac{1}{2}\frac{Q^2}{C}$$

P69:

Solution:

(a) The voltage across the terminals of the battery is

$$
\begin{aligned}
\Delta V_{bat} &= \text{emf} - I r_{int} \\
6.7 \text{ V} &= 9 \text{ V} - Ir
\end{aligned}
$$

To get the current through the battery, apply Ohm's Law to the resistor. The voltage across the resistor is 6.7 V.

$$
\begin{aligned}
\Delta V &= IR \\
I &= \frac{6.7 \text{ V}}{100 \text{ } \Omega} \\
&= 0.067 \text{ A}
\end{aligned}
$$

Solve for the internal resistance of the battery.

$$
\begin{aligned}
6.7 \text{ V} &= 9 \text{ V} - Ir \\
r &= \frac{(9 \text{ V} - 6.7 \text{ V})}{0.067 \text{ A}} \\
&= 34 \text{ } \Omega
\end{aligned}
$$

(b) If you connect a low-resistance wire across the terminals of the battery and connect the voltmeter across the same terminals, then the voltmeter leads will are "shorted," meaning that they are connected together with no resistance between them. According to Ohm's law, the potential difference is zero. The voltmeter will read zero.

(c) The only resistance is the internal resistance of the battery. Applying Ohm's law gives

$$
\begin{aligned}
0 &= \text{emf} - I r_{int} \\
I &= \frac{9 \text{ V}}{34 \text{ } \Omega} \\
&= 0.26 \text{ A}
\end{aligned}
$$

P73:

Solution:

Find the equivalent resistance of the voltmeter and the two resistors. The voltmeter is in parallel with the first resistor, $R_1 = 4 \times 10^6 \text{ } \Omega$. The equivalent resistance of the voltmeter and R_1 is

$$
\begin{aligned}
\frac{1}{R_{eq}} &= \frac{1}{4 \times 10^6 \text{ } \Omega} + \frac{1}{1 \times 10^6 \text{ } \Omega} \\
R_{eq} &= 8 \times 10^5 \text{ } \Omega
\end{aligned}
$$

This resistance is in series with the second resistor, R_2. Thus the equivalent resistance of the voltmeter and both resistors is:

$$R_{eq,circuit} = 8 \times 10^5 \ \Omega + 4 \times 10^6 \ \Omega$$
$$= 4.8 \times 10^6 \ \Omega$$

Apply Ohm's Law to the equivalent resistance of the voltmeter and both resistors to get the current through the battery.

$$I = \frac{\Delta V_{bat}}{R_{eq,circuit}}$$
$$= \frac{60 \ V}{4.8 \times 10^6 \ \Omega}$$
$$= 1.25 \times 10^{-5} \ A$$

The voltage across the voltmeter and R_1 is:

$$\Delta V_1 = IR_{eq}$$
$$= (1.25 \times 10^{-5} \ A)(8 \times 10^5 \ \Omega)$$
$$= 10 \ V$$

A quick way to check your answer is to calculate the voltage across R_2. The voltages across R_1 and R_2 should add to 60 V, so we expect the voltage across R_2 to be 50 V. Let's see.

$$\Delta V_2 = IR_2$$
$$= (1.25 \times 10^{-5} \ A)(4 \times 10^6 \ \Omega)$$
$$= 50 \ V \qquad \text{exactly as expected}$$

P77:
  Solution:

(a) The minimum electric field needed to ionize air is $3 \times 10^6 \ \frac{V}{m}$. The electric field inside the capacitor is

$$\Delta V_c = Es$$
$$E = \frac{\Delta V_c}{s} = \frac{9 \ V}{2.5 \times 10^3 \ m}$$
$$= 3600 \ \frac{V}{m}$$

This is less than the critical field needed to produce a spark.

(b) $Q = C\Delta V$. The capacitance of the capacitor is

$$C = \frac{\epsilon_0 A}{s} = \frac{(8.85 \times 10^{-12} \ \frac{F}{m})(4 \ m)(3 \ m)}{2.5 \times 10^{-3} \ m}$$
$$= 4.25 \times 10^{-8} \ F$$

$$Q = (4.25 \times 10^{-8} \text{ F})(9 \text{ V})$$
$$= 3.83 \times 10^{-7} \text{ C}$$

(c) Inserting the plastic changes ΔV across the plates because E in the plastic is less than before. Total ΔV across the plates is the sum of ΔV across the plastic + ΔV across the air gap.

$$
\begin{aligned}
\Delta V &= \Delta V_{\text{plastic}} + \Delta V_{\text{air}} \\
&= \frac{(9 \text{ V})(\frac{1 \text{ mm}}{2.5 \text{ mm}})}{K} + (9 \text{ V})\left(\frac{1.5 \text{ mm}}{2.5 \text{ mm}}\right) \\
&= 0.72 \text{ V} + 5.4 \text{ V} \\
&= 6.12 \text{ V}
\end{aligned}
$$

(d) The voltmeter and capacitor make an RC circuit. The capacitor will discharge. ΔV is

$$\Delta V = (6.12 \text{ V})e^{-\frac{t}{RC}}$$

where

$$
\begin{aligned}
RC &= (100 \times 10^6 \ \Omega)(4.25 \times 10^{-8} \text{ F}) \\
&= 4.25 \text{ s}
\end{aligned}
$$

$$
\begin{aligned}
\Delta V &= (6.12 \text{ V})e^{-\frac{3 \text{ s}}{4.25 \text{ s}}} \\
&= 3.02 \text{ V}
\end{aligned}
$$

P79:

Solution:

(a) Loop 1: ABCEFGA

$$20 - 10I_1 - 15I_4 - 12I_6 - 20I_1 = 0$$

Loop 2: DECD

$$5 + 15I_4 - 20I_2 = 0$$

Loop 3: DFED

$$-30I_3 + 12I_6 - 5 = 0$$

Node rule at C:

$$I_1 = I_2 + I_4$$

Node rule at D:

$$I_2 = I_5 + I_3$$

Node rule at F:

$$I_3 + I_6 = I_1$$

Solving these equations using linear algebra or the solve function on your calculator (which uses linear algebra) gives

$$I_1 = 0.439 \text{ A}$$
$$I_2 = 0.331 \text{ A}$$
$$I_3 = 0.00649 \text{ A}$$
$$I_4 = 0.108 \text{ A}$$
$$I_5 = 0.325 \text{ A}$$
$$I_6 = 0.433 \text{ A}$$

(b) Check each equation:

$$20 - 10(0.439) - 15(0.108) - 12(0.433) - 20(0.439) \stackrel{?}{=} 0.014$$
$$5 + 15(0.108) - 20(0.331) \stackrel{?}{=} 0$$
$$-30(0.00649) + 12(0.433) - 5 \stackrel{?}{=} 0.0013$$
$$0.439 - 0.331 - 0.108 \stackrel{?}{=} 0$$
$$0.331 - 0.325 - 0.00649 \stackrel{?}{=} -0.00049$$
$$0.00649 + 0.433 - 0.439 \stackrel{?}{=} 0.00049$$

Results are very close to zero. Any difference from zero is due to rounding.

(c)

$$
\begin{aligned}
V_{\mathrm{C}} - V_{\mathrm{G}} &= (V_{\mathrm{A}} - V_{\mathrm{G}}) + (V_{\mathrm{B}} - V_{\mathrm{A}}) + (V_{\mathrm{C}} - V_{\mathrm{B}}) \\
&= 20 I_1 + 20\ \text{V} + 10 I_1 \\
&= 20(0.439) + 20\ \text{V} + 10(0.439) \\
&= 33.2\ \text{V}
\end{aligned}
$$

The voltmeter will read a positive potential difference.

(d)

$$
\begin{aligned}
P &= \Delta V I_5 \\
&= (5\ \text{V})(0.325\ \text{A}) \\
&= 1.63\ \text{W}
\end{aligned}
$$

(e)

$$
\begin{aligned}
\Delta V &= EL \\
E &= \frac{\Delta V}{L} = \frac{(12\ \Omega) I_6}{L} \\
&= \frac{(12\ \Omega)(0.433\ \text{A})}{0.003\ \text{m}} \\
&= 1730\ \frac{\text{V}}{\text{m}}
\end{aligned}
$$

20   Chapter 20: Magnetic Force

Q03:
Solution:

(a) The proton's $\vec{E}$ is to the right. There is no magnetic field.

(b) The electric force on the electron is to the left. There is no magnetic force.

(c) The electron's $\vec{E}$ is to the right. The electron's magnetic field is into the page.

(d) The electric force is to the right. The magnetic force is up.

(e) Reciprocity apparently does not apply to magnetic forces.

(f) If reciprocity does not apply, then the total system momentum will change. The momentum principle is violated **unless** as assign momentum to the fields.

Q09:
Solution:

Conventional current flows in the $-y$ direction through the resistor. The magnetic force on mobile electrons in the bar is in the $-y$ direction which causes the electron "sea" to shift downward, leaving the top end of the bar positively charged and the bottom end of the bar negatively charged. However, because the ends are connected by a conductor, positive charge (conventional current) will flow from the top end, through the resistor, and to the bottom end of the bar. As a result, conventional current flows counterclockwise through the circuit.

Since I flows in the $+y$ direction and $\vec{B}$ is in the $-z$ direction, the magnetic force on the wire, $\vec{F}_{mag} = I\vec{L} \times \vec{B}$, is in the $-x$ direction. If you pull the bar at a constant speed v in the $+x$ direction, you must exert a force $\vec{F}$ in the $+x$ direction that is equal in magnitude to the magnetic force in the $-x$ direction so that the net force on the bar is zero.

Q13:
Solution:

The bar magnet initially rotates clockwise. The magnetic dipole moment points from S to N, so $\vec{B}$ and $\vec{\mu}$ are in the directions shown below.

The torque on the bar magnet is $\vec{\tau} = \vec{\mu} \times \vec{B}$ which is in the -z direction. Since $\vec{\tau} \propto \Delta \vec{L}$ and since $\vec{L}_i = 0$ then the final angular momentum after a time Δt is in the -z direction. The bar magnet rotates clockwise.

Q21:
Solution:

The energy needed to ionize the molecule is proportional to $\Delta V \approx 14V$. Since $\Delta V = E_{crit}d$, then if it takes $6E_0$ to accelerate an electron to sufficient energy to ionize a molecule, then the electron must travel $1/6$ the distance. Thus $\frac{d}{d_0} = 1/6$ where d_0 is the mean free path at STP.

P25:
Solution:

$$
\begin{aligned}
\vec{F} &= q\vec{v} \times \vec{B} \\
&= (1.6 \times 10^{-19} \text{ C})(\langle -5 \times 10^5, 0, 0 \rangle \text{ m/s}) \times \langle 0, 0, 3.5 \rangle \text{ T} \\
&= \langle 0, 0, 1.92 \times 10^{-14} \rangle \text{ N} \\
\left| \vec{F} \right| &= 1.92 \times 10^{-14} \text{ N} \\
\hat{F} &= <0, 0, 1>
\end{aligned}
$$

P29:
Solution:

$\vec{F}$ is toward the center of the circle. The right-hand rule shows that $\vec{B}$ is outward, in the $+z$ direction, so $\hat{B} = <0, 0, 1>$.

$$
\begin{aligned}
\left| \vec{F}_{net} \right| &= \left| \frac{d\vec{p}}{dt} \right| \\
\left| \vec{F}_{mag} \right| &= \left| \frac{d\vec{p}}{dt} \right| \\
qvB &= \frac{mv^2}{R} \\
qB &= \frac{mv}{R} \\
B &= \frac{mv}{qR} \\
&= \frac{m \left(\frac{2\pi R}{T} \right)}{qR} \\
&= \frac{m2\pi}{qT} \\
&= \frac{4(1.67 \times 10^{-27} \text{ kg})2\pi}{2(1.6 \times 10^{-19} \text{ C})(80 \times 10^{-9} \text{ s})} \\
&= 1.64 \text{ T}
\end{aligned}
$$

P33:
Solution:

(a) The needle deflects inward.

$$\left|\vec{B}\right| = \frac{\mu_o}{4\pi} \frac{2N\pi I}{R} \text{into page}$$

$$\tan\theta = \frac{\left|\vec{B}\right|}{\left|\vec{B}_{\text{Earth}}\right|}$$

$$= \frac{\frac{\left(1\times10^{-7} \frac{\text{T}\cdot\text{m}^2}{\text{C}\cdot\text{m/s}}\right)(2)(3)\pi\frac{1.5\,\text{V}}{6\,\Omega}}{\left(7.5\times10^{-2}\,\text{m}\right)}}{\left(4\times10^{-5}\,\text{T}\right)}$$

$$\approx 0.1571$$

$$\therefore \theta \approx 8.9°$$

(b)

$$\left|\vec{F}_{\text{mag}}\right| = \left|-e\vec{v}\times\vec{B}\right| = 0$$

$$\left|\vec{F}_{\text{el}}\right| = \left|-e\vec{E}\right| \text{ down}$$

$$\left|\vec{F}_{\text{el}}\right| = \left(1.6\times10^{-19}\,\text{C}\right)(250\,\text{N/C})$$

$$\approx 4\times10^{-17}\,\text{N}$$

P37:
Solution:

$$\vec{F}_{\text{mag}} = I\vec{L}\times\vec{B}$$

$$= (1.8\,\text{A})(\langle 0.25, 0, 0\rangle\,\text{m})\times\langle 0, 0.54, 0\rangle\,\text{T}$$

$$= \langle 0, 0, 0.243\rangle\,\text{N}$$

$$\left|\vec{F}\right| = 0.243\,\text{N}$$

$$\hat{F} = <0, 0, 1>$$

P43:
Solution:

The metal rod is in equilibrium, so $\vec{F}_{\text{net}} = 0$. The gravitational force by Earth on the rod is downward, so the magnetic force by the magnetic field on the rod must be upward. Since $\vec{F}_{\text{mag}} = I\vec{L}\times\vec{B}$ and I flows in the +x direction, $\vec{B}$ must be in the -z direction. Applying the Momentum Principle gives

$$\begin{aligned}
\vec{F}_{net} &= 0 \\
\vec{F}_{mag} + \vec{F}_{grav} &= 0 \\
\left|\vec{F}_{mag}\right| &= \left|\vec{F}_{grav}\right| \\
ILB &= mg \\
B &= \frac{mg}{IL} \\
&= \frac{(0.07 \text{ kg})(9.8 \frac{\text{N}}{\text{kg}})}{(5 \text{ A})(0.12 \text{ m})} \\
&= 1.14 \text{ T}
\end{aligned}$$

So, $\vec{B} = \langle 0, 0, -1.14 \rangle$ T.

P45:
Solution:

Since $\vec{F}_{net} = 0$, then

$$\begin{aligned}
\left|\vec{F}_{elec}\right| &= \left|\vec{F}_{mag}\right| \\
qE &= qvB \\
v &= \frac{E}{B} \\
&= \frac{3800 \frac{\text{V}}{\text{m}}}{0.4 \text{ T}} \\
&= 9500 \text{ m/s}
\end{aligned}$$

P51:
Solution:

Define +x to the left, +y upward, and +z into the page, in accordance with a right-hand coordinate system.

(a) Electrons are accelerated to the left, so B is the positively charged plate.

(b) $\vec{F}_{elec}$ is downward. Since $\vec{F} = q\vec{E}$ and q is negative, then $\vec{E}$ is upward.

(c) $\vec{F}_{mag}$ is upward, as determined by $\vec{F}_{mag} = q\vec{v} \times \vec{B}$.

(d) Use Conservation of Energy for the electron and accelerating plates.

$$\begin{aligned}
\Delta E_{\text{sys}} &= 0 \\
\Delta K + \Delta U &= 0 \\
K_f - \cancelto{0}{K_i} + q\Delta V &= 0 \\
\frac{1}{2}mv_f^2 &= -q\Delta V \\
v_f &= \left(\frac{-2(-1.6 \times 10^{-19}\text{ C})(3.1 \times 10^3\text{ V})}{9.11 \times 10^{-31}\text{ kg}}\right)^{\frac{1}{2}} \\
&= 3.30 \times 10^7\text{ m/s}
\end{aligned}$$

Note that the classical approximation $K \approx \frac{1}{2}mv^2$ was used, and v_f came out to $0.1c$. If $K = (\gamma - 1)mc^2$ is used, then $v_f = 3.39 \times 10^7$ m/s. The relativistically correct equation for kinetic energy should be used for a more accurate solution. The classical approximation results in roughly 3% error.

The magnetic field between the coils has a magnitude

$$\begin{aligned}
B &= 2\left(\frac{\mu_o}{4\pi}\frac{2\pi R^2 I N}{(z^2 + R^2)^{\frac{3}{2}}}\right) \\
&= 2(1 \times 10^{-7}\frac{\text{T}\cdot\text{m}}{\text{A}})\frac{2\pi(0.06\text{ m})^2(0.5\text{ A})(320)}{((0.03\text{ m})^2 + (0.06\text{ m})^2)^{\frac{3}{2}}} \\
&= 2.40 \times 10^{-3}\text{ T}
\end{aligned}$$

(e) (2), (5), and (6) are true.

(f) $\vec{F}_{\text{net}} = 0$, and $\Delta V_{\text{def}} = Ed$ where d is the distance between the deflecting plates. So,

$$\begin{aligned}
F_{\text{elec}} &= F_{\text{mag}} \\
qE &= qvB \\
E &= vB \\
\frac{\Delta V}{d} &= vB \\
\Delta V &= vBd \\
&= (3.39 \times 10^7\text{ m/s})(2.4 \times 10^{-3}\text{ T})(0.008\text{ m}) \\
&= 650\text{ V}
\end{aligned}$$

P55:
  Solution:

  (a) A sketch of the bar is shown in the figure below.

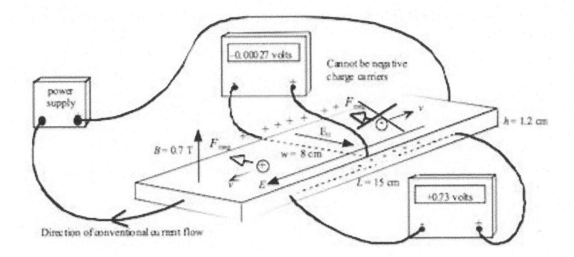

The voltmeter reading of -0.00027 volts indicates that the back side of the slab is at a higher potential than the front side, so there is a Hall-effect transverse electric field E_H as shown. That means that there must be extra $+$ charge on the back side (and extra $-$ charge on the front).

The conventional current flow and the main electric field E point as shown, and the $+0.73$ volt reading is consistent: the potential is dropping in the direction of E and the conventional current. Positive carriers move in the direction of conventional current, and experience a magnetic force toward the backside, which would lead to $+$ charge buildup on the back side, which is what is observed.

Negative carriers move opposite to the conventional current, and would also experience a magnetic force toward the back side, which would lead to $-$ charge buildup on the back side, which is not observed.

So the charge carriers are positive.

(b) In the steady state, the transverse electric and magnetic forces must balance, so $qE_H = qvB\sin(90°) = qvB$. So $E_H = vB$. The drift speed v is uniform throughout the slab (current conservation and constant cross-sectional area), and the magnetic field B is uniform throughout this region. So $E_H = vB$ must be uniform along the 8-cm path across the slab, and we can write

$$
\begin{aligned}
|\Delta V| &= 0.00027 \text{ V} \\
&= E_H w \\
&= E_H (0.08 \text{ m}) \\
E_H &= \frac{0.00027 \text{ V}}{0.08 \text{ m}} \\
&= 0.0034 \text{ V/m}
\end{aligned}
$$

$$
\begin{aligned}
v &= \frac{E_H}{B} \\
&= \frac{0.0034 \text{ V/m}}{0.7 \text{ T}} \\
&= 4.8 \times 10^{-3} \text{ m/s}
\end{aligned}
$$

Note that we have experimentally determined the drift speed v, independent of the carrier charge q and the density of charge carriers n.

(c) $v = uE$, where E is the electric field in the direction of the current. Since v is uniform, E must be uniform, and we can write

$$
\begin{aligned}
|\Delta V| &= EL \\
E &= \frac{0.73 \text{ V}}{0.15 \text{ m}} \\
&= 4.9 \text{ V/m}
\end{aligned}
$$

Note that the electric field E in the direction of the current is much larger than the transverse field E_H. The mobility is

$$
\begin{aligned}
u &= \frac{v}{E} \\
&= \frac{4.8 \times 10^{-3} \text{ m/s}}{4.9 \text{ V/m}} \\
&= 9.8 \times 10^{-4} \frac{\text{m/s}}{\text{V/m}}
\end{aligned}
$$

(d) $I = qnAv$, and we assume that $q = e$. So we have

$$
\begin{aligned}
n &= \frac{I}{eAv} \\
&= \frac{0.3 \text{ A}}{(1.602 \times 10^{-19} \text{ C})(0.08 \text{ m})(0.012 \text{ m})(4.8 \times 10^{-3} \text{ m/s})} \\
&= 4 \times 10^{23} \text{ m}^{-3}
\end{aligned}
$$

This is a very low density of charge carriers. The density of free electrons in copper is 8×10^{28} m^{-3}. Evidently this slab of material is not an ordinary metal: the charge carriers are positive, and the density of charge carriers is very low.

(e) The potential difference along 15 cm of material is 0.73 V. So,

$$
\begin{aligned}
\Delta V &= IR \\
R &= \frac{0.73 \text{ V}}{0.3 \text{ A}} \\
&= 2.4 \ \Omega
\end{aligned}
$$

P61:
   Solution:

(2) According to the right-hand rule, the magnetic force on the negatively charged mobile electrons is to the left, causing negative charge to pile up on the left end of the bar and positive charge to pile up on the right end of the bar.

P65:
   Solution:

See the figure below for a pictorial representation of the charge distribution and applied force.

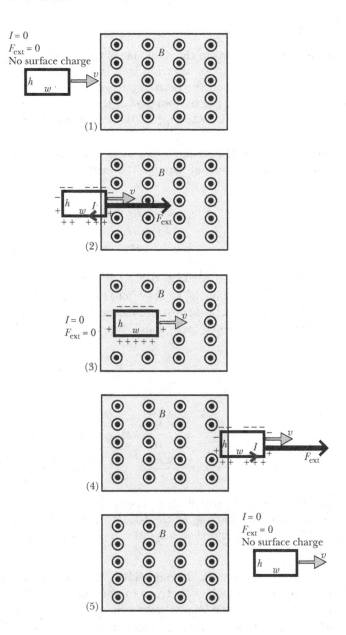

(a)   • The magnetic force on mobile electrons in the wire loop is zero, so there is no potential difference across any part of the wire loop and no current flows in the loop.

$$I \;\; = \;\; 0$$

• The magnetic force on the right side of the loop is $\vec{F}_{mag} = IhB$. Since $B = 0$, the magnetic force on the right side of the loop is zero. The same is true of every part of the loop. So, the applied force needed to keep the loop moving at constant velocity is zero.

$$F_{applied} \;\; = \;\; 0$$

- There is no surface charge on the wire loop.

(b)   • The magnetic force on mobile electrons in the right side of the wire loop is given by $\vec{F}_{mag} = q\vec{v} \times \vec{B}$. According to the right-hand rule, the force on mobile electrons is in the $+y$ direction, so electrons pile up on the top part of the loop. Positive charge piles up on the bottom part of the loop. Mobile electrons will flow from the top part of the loop, counterclockwise, to the bottom part of the loop, and the magnetic force in the right, vertical piece of the loop will push them back to the top of the loop again. Electron current flows counterclockwise, so conventional current flows clockwise.

    The potential difference across the right-side of the loop has a magnitude $|\Delta V| = vBh$ and is equal to IR, according to Ohm's law. Thus, the current through the loop is

$$\begin{aligned} IR &= vBh \\ I &= \frac{vBh}{R} \quad \text{clockwise} \end{aligned}$$

- The magnetic force on the right side of the loop is $\vec{F}_{mag} = IhB$. The direction of the force on the right-side is $\vec{F}_{mag} = I\vec{L} \times \vec{B}$. The vector $\vec{L}$ points in the $-y$ direction, in the direction of conventional current on the right side of the loop. According to the right-hand rule, the magnetic force is thus in the $-x$ direction. To move at constant velocity, the applied force (by your hand, for example) must be to the right, in the $+x$ direction, with an equal magnitude.

$$\begin{aligned} F_{applied} &= F_{mag} \\ &= IhB \\ &= \frac{vBh}{R}(hB) \\ &= \frac{vB^2h^2}{R} \quad \text{in the } +x \text{ direction} \end{aligned}$$

- The surface charge is negative on the top of the right-side and positive on the bottom of the right side. Going counterclockwise around the loop starting at the top, it decreases in negative charge density until going to zero (at the middle of the left side) and then increases in positive charge density from the middle, left side to the bottom, right side.

(c)   • The magnetic force on mobile electrons in the right side of the wire loop is given by $\vec{F}_{mag} = q\vec{v} \times \vec{B}$. According to the right-hand rule, the force on mobile electrons **in the right side of the loop** is in the $+y$ direction, so electrons pile up on the top part of the loop. Positive charge piles up on the bottom part of the loop.

    In the left side of the loop, there is also a magnetic force upward on the mobile electrons causing negative charge to build up on the top of the left side and positive charge to build up on the bottom of the left side. There is no surface charge gradient around the loop; therefore, no current will flow through the loop.

$$I = 0$$

- The magnetic force on the right side of the loop is $\vec{F}_{mag} = IhB$. Since $I = 0$, the magnetic force on the right side of the loop is zero. The same is true of every other part of the loop. So, the applied force needed to keep the loop moving at constant velocity is zero.

$$F_{applied} = 0$$

- There is negative surface charge on the top of the wire loop and positive surface charge on the bottom of the loop. It is uniform charge density on the top and uniform charge density on the bottom. Since there is no gradient (from right to left, for example), no current flows through the loop.

(d)

- The magnetic force on mobile electrons in the right side of the wire loop is zero because it is outside the magnetic field. The magnetic force on mobile electrons in the left side of the wire loop is given by $\vec{F}_{mag} = q\vec{v} \times \vec{B}$. According to the right-hand rule, the force on mobile electrons is in the $+y$ direction, so electrons pile up on the top part of the loop. Positive charge piles up on the bottom part of the loop. Mobile electrons will flow from the top part of the loop on the left, clockwise, to the bottom part of the loop on the left, and the magnetic force in the left, vertical piece of the loop will push them back to the top of the loop again. Electron current flows clockwise, so conventional current flows counterclockwise.

  The potential difference across the left-side of the loop has a magnitude $|\Delta V| = vBh$ and is equal to IR, according to Ohm's law. Thus, the current through the loop is

$$
\begin{aligned}
IR &= vBh \\
I &= \frac{vBh}{R} \qquad \text{counterclockwise}
\end{aligned}
$$

- The magnetic force on the left side of the loop is $\vec{F}_{mag} = IhB$. The direction of the force on the left-side is $\vec{F}_{mag} = I\vec{L} \times \vec{B}$. The vector $\vec{L}$ points in the $-y$ direction, in the direction of conventional current on the left side of the loop. According to the right-hand rule, the magnetic force is in the $-x$ direction. To move at constant velocity, the applied force (by your hand, for example) must be to the right, in the $+x$ direction, with an equal magnitude.

$$
\begin{aligned}
F_{applied} &= F_{mag} \\
&= IhB \\
&= \frac{vBh}{R}(hB) \\
&= \frac{vB^2h^2}{R} \qquad \text{in the } +x \text{ direction}
\end{aligned}
$$

- The surface charge is negative on the top of the left side and positive on the bottom of the left side. Going clockwise around the loop starting at the top, it decreases in negative charge density until going to zero (at the middle of the right side) and then increases in positive charge density from the middle, right side to the bottom, left side.

(e)

- The magnetic force on mobile electrons in the wire loop is zero, so there is no potential difference across any part of the wire loop and no current flows in the loop.

$$
I = 0
$$

- The magnetic force on the left or right side of the loop is $\vec{F}_{mag} = IhB$. Since $B = 0$, the magnetic force on the left or right side of the loop is zero. The same is true of every part of the loop. So, the applied force needed to keep the loop moving at constant velocity is zero.

$$
F_{applied} = 0
$$

- There is no surface charge on the wire loop.

P69:

 Solution:

(a) The magnetic force on mobile electrons in the right side of the wire loop is zero because it is outside the magnetic field. The magnetic force on mobile electrons in the left side of the wire loop is given by $\vec{F}_{mag} = q\vec{v} \times \vec{B}$. According to the right-hand rule, the force on mobile electrons is in the $-y$ direction, so electrons pile up on the bottom part of the loop. Positive charge piles up on the top part of the loop. Mobile electrons will flow from the bottom part of the loop on the left, counterclockwise, to the top part of the loop on the left, and the magnetic force in the left, vertical piece of the loop will push them back to the bottom of the loop again. Electron current flows counterclockwise, so conventional current flows clockwise.

The surface charge is negative on the bottom of the left side and positive on the top of the left side. Going counterclockwise around the loop starting at the bottom, it decreases in negative charge density until going to zero (at the middle of the right side) and then increases in positive charge density from the middle, right side to the top, left side.

(b) The emf across the left-side of the loop has a magnitude emf $= vBh$ and is equal to IR, according to Ohm's law. Thus, the current through the loop is

$$
\begin{aligned}
IR &= vBh \\
I &= \frac{vBh}{R} \\
&= \frac{(8 \text{ m/s})(1.2 \text{ T})(0.03 \text{ m})}{0.3 \ \Omega} \\
&= 0.96 \text{ A}
\end{aligned}
$$

Conventional current flows clockwise, as described in part (a) above.

(c) **(2) is true.** Note that (1) is false; though the magnetic force does stretch the loop vertically, there is a net magnetic force to the left on the loop. (3) is false because the magnetic force on the loop is to the left.

(d) The magnetic force on the left side of the loop has a magnitude $\left|\vec{F}_{mag}\right| = IhB$. The direction of the force on the left-side is given by $\vec{F}_{mag} = I\vec{L} \times \vec{B}$. The vector $\vec{L}$ points in the $+y$ direction, in the direction of conventional current on the left side of the loop. According to the right-hand rule, the magnetic force is in the $-x$ direction. To move at constant velocity, the applied force (by your hand, for example) must be to the right, in the $+x$ direction, with an equal magnitude.

$$
\begin{aligned}
F_{applied} &= F_{mag} \\
&= IhB \\
&= (0.96 \text{ A})(0.03 \text{ m})(1.2 \text{ T}) \\
&= 0.035 \text{ N}
\end{aligned}
$$

P71:

 Solution:

$$
\begin{aligned}
\vec{\tau} &= \vec{\mu} \times \vec{B} \\
&= \langle 4, 0, 1.5 \rangle \text{ A} \cdot \text{m}^2 \times \langle 0.8, 0, 0 \rangle \text{ T} \\
&= \langle 0, 1.2, 0 \rangle \text{ N} \cdot \text{m}
\end{aligned}
$$

P77:

  Solution:

(a) Since the entire atom is neutral, then the "atomic core" has a charge $q = +e$.

(b) Treat the core as a point particle, then

$$V = \frac{1}{4\pi\varepsilon_o}\frac{q}{r}$$

$$\approx 9 \times 10^9 \, \frac{\text{N}\cdot\text{m}^2}{\text{C}^2} \frac{1.602 \times 10^{-19}\,\text{C}}{(1 \times 10^{-10}\,\text{m})}$$

$$\approx 14.4 \text{ V}$$

(c)

$$\lim_{r\to\infty} V = 0$$

(d) The potential difference is $V_\infty - V_r = -14.4$ V

(e) The change in potential energy of the system in moving an electron from $r = 1 \times 10^{-10}$ m to a distance far away is $\Delta U = q_{electron}\Delta V_{core} = (e)(-14.4 \text{ V}) = 14.4$ eV.

(f) A free incoming particle that collides with the atom must have a kinetic energy of at least 14.4 eV in order to "knock" an electron from the atom and thereby ionize the atom.

CP81:

  Solution:

(a) Here is the program.

```
from __future__ import division, print_function
from visual import *

scene.width = 800
scene.height = 800
## CONSTANTS ##
mzofp = 1e-7
qe = 1.6e-19
mproton = 1.7e-27
B0 = vector(0,0.2,0)
bscale = 1
#### THIS CODE DRAWS A GRID ##
#### AND DISPLAYS MAGNETIC FIELD ##
xmax = 0.4
dx = 0.1
yg = -0.1
x = -xmax
while x < xmax+dx:
    curve(pos=[(x,yg,-xmax),(x,yg,xmax)],
        color=(0.7,0.7,0.7))
    x = x + dx
```

```
z = −xmax
while z < xmax+dx:
    curve(pos=[(−xmax,yg,z),(xmax,yg,z)],
          color=(0.7,0.7,0.7))
    z = z + dx
x = −xmax
dx = 0.2
while x < xmax+dx:
    z = −xmax
    while z < xmax+dx:
        arrow(pos=(x,yg,z),
              axis=B0*bscale,
              color=(0,0.8,0.8))
        z = z + dx
    x = x + dx


#### OBJECTS AND INITIAL CONDITIONS ##
particle = sphere(pos=vector(0,0.15,0.3),
                  radius=1e−2,
                  color=color.yellow,
                  make_trail=True)
## make trail easier to see (thicker) ##
particle.trail_object.radius = particle.radius/3
vparticle = vector(−2e6,0,0)
p = mproton*vparticle
qparticle = qe
deltat = 5e−11
t=0
###############################################
while t < 3.34e−7 :
    rate(500)
    Fnet = qparticle*cross(vparticle,B0)
    p = p + Fnet * deltat
    vparticle = p/mproton
    particle.pos = particle.pos + (p/mproton) * deltat
    t = t + deltat
```

(b) You can calculate the radius by equating the magnitude of the Lorentz force on the particle to the magnitude of the centripetal force necessary for circular motion.

$$q\,|\vec{v}|\,\left|\vec{B}\right| = \frac{m\,|\vec{v}|^2}{r}$$

$$r = \frac{m\,|\vec{v}|}{q\,\left|\vec{B}\right|}$$

$$= \frac{(1.7 \times 10^{-27}\ \text{kg})(2 \times 10^{6}\ \text{m/s})}{(1.6 \times 10^{-19}\ \text{C})(0.2\ \text{T})}$$

$$\approx 0.106\ \text{m}$$

(c) The trajectory's radius is proportional to the particle's speed so doubling the speed doubles the radius.

(d) The orbital period is independent of speed, so doubling the speed causes no change in the orbital period.

$$|\vec{v}| = \frac{2\pi r}{T}$$

$$T = \frac{2\pi r}{|\vec{v}|} = \frac{2\pi m\,|\vec{v}|}{q\left|\vec{B}\right||\vec{v}|}$$

$$T = \frac{2\pi m}{q\left|\vec{B}\right|}$$

(e) Again, radius is proportional to speed so halving the speed halves the radius.

(f) Again, orbital period is independent of speed, so there is no change.

(g) Radius is inversely proportional to field magnitude, so doubling the field halves the radius.

(h) The particle's motion is not affected by the magnetic field at all.

(i) The particle's trajectory becomes a helix.

21   Chapter 21: Patterns of Field in Space

Q05:

Solution:

According to Gauss' law for magnetism, the net magnetic flux through a closed surface must be zero. However, in this case, there is a net flux through the surface because the flux out of the bottom end of the cylinder is greater than the flux into the top end of the cylinder. As a result, the measurements of the magnetic field must be incorrect.

P07:

Solution:

$\Phi_{front} = \Phi_{back} = \Phi_{top} = \Phi_{bottom} = 0$ since $\vec{E}$ is parallel to $\hat{n}$.

$$
\begin{aligned}
\Phi_{left} &= \vec{E} \cdot \hat{n} A \\
&= <400\,\frac{V}{m}, 0, 0> \, \cdot \, <-1, 0, 0> (0.03\text{ m})(0.02\text{ m}) \\
&= -0.24\text{ V} \cdot \text{m}
\end{aligned}
$$

$$
\begin{aligned}
\Phi_{right} &= \langle 1000, 0, 0 \rangle\,\frac{V}{m} \cdot \, <1, 0, 0> (0.03\text{ m})(0.02\text{ m}) \\
&= 0.6\text{ V} \cdot \text{m}
\end{aligned}
$$

$$
\begin{aligned}
\Phi_{total} &= \Sigma_{surface}\,\Phi \\
&= 0.6\text{ V} \cdot \text{m} + -0.24\text{ V} \cdot \text{m} \\
&= 0.36\text{ V} \cdot \text{m}
\end{aligned}
$$

Gauss Law:

$$
\begin{aligned}
\Phi_{total} &= \frac{\Sigma q_{inside}}{\epsilon_0} \\
\Sigma q_{inside} &= \epsilon_0\,(0.36\text{ V} \cdot \text{m}) \\
&= 3.19 \times 10^{-12}\text{ C}
\end{aligned}
$$

P13:

Solution:

(a)

$$\begin{aligned}
\hat{n}_{\text{top}} &= \ <0,1,0> \\
\hat{n}_{\text{bottom}} &= \ <0,-1,0> \\
\hat{n}_{\text{right}} &= \ <1,0,0> \\
\hat{n}_{\text{left}} &= \ <-1,0,0> \\
\hat{n}_{\text{front}} &= \ <0,0,1> \\
\hat{n}_{\text{back}} &= \ <0,0,-1>
\end{aligned}$$

(b)

$$\begin{aligned}
\Phi_{\text{top}} &= \vec{E}_{\text{top}} \cdot \hat{n}_{\text{top}} A_{\text{top}} \\
&= \vec{E}_{2} \cdot \ <0,1,0> (wd) \\
&= \langle 120,96,0 \rangle \ \frac{\text{N}}{\text{C}} \cdot \ <0,1,0> (0.06\text{ m})(0.02\text{ m}) \\
&= (96\ \frac{\text{N}}{\text{C}})(0.06\text{ m})(0.02\text{ m}) \\
&= 0.1152\ \frac{\text{N} \cdot \text{m}^2}{\text{C}}
\end{aligned}$$

$$\begin{aligned}
\Phi_{\text{bottom}} &= \vec{E}_{\text{bottom}} \cdot \hat{n}_{\text{bottom}} A_{\text{bottom}} \\
&= \langle 50,40,0 \rangle \ \frac{\text{N}}{\text{C}} \cdot \ <0,-1,0> (0.06\text{ m})(0.02\text{ m}) \\
&= (-40\ \frac{\text{N}}{\text{C}})(0.06\text{ m})(0.02\text{ m}) \\
&= -0.048\ \frac{\text{N} \cdot \text{m}^2}{\text{C}}
\end{aligned}$$

$$\begin{aligned}
\Phi_{\text{right}} &= \vec{E}_{\text{right}} \cdot \hat{n}_{\text{right}} A_{\text{right}} \\
&= \langle 120,96,0 \rangle \ \frac{\text{N}}{\text{C}} \cdot \ <1,0,0> (0.04\text{ m})(0.02\text{ m}) \\
&= (120\ \frac{\text{N}}{\text{C}})(0.04\text{ m})(0.02\text{ m}) \\
&= 0.096\ \frac{\text{N} \cdot \text{m}^2}{\text{C}}
\end{aligned}$$

$$\begin{aligned}
\Phi_{\text{left}} &= \vec{E}_{\text{left}} \cdot \hat{n}_{\text{left}} A_{\text{left}} \\
&= \langle 120,96,0 \rangle \ \frac{\text{N}}{\text{C}} \cdot \ <-1,0,0> (0.04\text{ m})(0.02\text{ m}) \\
&= -0.096\ \frac{\text{N} \cdot \text{m}^2}{\text{C}}
\end{aligned}$$

$$\Phi_{\text{front}} = \vec{E}_{\text{front}} \cdot \hat{n}_{\text{front}} A_{\text{front}}$$
$$= (\langle 120, 96, 0 \rangle \frac{\text{N}}{\text{C}}) \cdot <0, 0, 1> (0.06 \text{ m})(0.04 \text{ m})$$
$$= 0$$

$$\Phi_{\text{back}} = 0$$

(c)

$$\Phi_{\text{total}} = \Sigma \Phi_{\text{side}}$$
$$= \Phi_{\text{top}} + \Phi_{\text{bottom}}$$
$$= 0.1152 \frac{\text{N} \cdot \text{m}^2}{\text{C}} + -0.048 \frac{\text{N} \cdot \text{m}^2}{\text{C}}$$
$$= 0.0672 \frac{\text{N} \cdot \text{m}^2}{\text{C}}$$

According to Gauss' Law,

$$\oint \vec{E} \cdot \hat{n} A = \frac{\Sigma q_{\text{inside}}}{\epsilon_0}$$
$$\Sigma q_{\text{inside}} = (0.0672 \frac{\text{N} \cdot \text{m}^2}{\text{C}})(8.85 \times 10^{-12} \frac{\text{C}^2}{\text{N} \cdot \text{m}^2})$$
$$= 5.95 \times 10^{-13} \text{ C}$$

P19:
Solution:

(a) $\vec{E}$ points to the right inside the capacitor. $\vec{E} = 0$ inside the metal plates. For Gaussian surface 1, it has 6 sides. For each side, $\Phi_{\text{side}} = \vec{E}_{\text{side}} \cdot \hat{n}_{\text{side}} A_{\text{side}}$.

For the top, bottom, front, and back, $\vec{E}$ is $\perp$ to $\hat{n}$, so $\Phi = 0$, since $\vec{E} \cdot \hat{n} = 0$.

For the left side, $E = 0$ (inside the metal), so $\Phi = 0$. As a result, only the right side has non-zero flux through it.

For the right side, $\hat{n} = <1, 0, 0>$ and $\vec{E} = <E, 0, 0>$. Thus,

$$\Phi_1 = \vec{E} \cdot <1, 0, 0> A_{\text{box}}$$
$$= E A_{\text{box}}$$

According to Gauss' Law for surface 1,

$$\Phi = \frac{\Sigma q_{inside}}{\epsilon_0}$$

$$EA_{box} = \frac{Q_{box}}{\epsilon_0}$$

Note that $\frac{Q_{box}}{A_{box}}$ is the same as $\frac{Q}{A}$ since the plate is uniformly charged. Thus,

$$E = \frac{\frac{Q}{A}}{\epsilon_0}$$

This agrees with Chapter 16 results.

(b) image goes here

The flux through the top, bottom, front, and back sides is zero since $\vec{E} \cdot \hat{n} = 0$ for these sides.

Apply Gauss' Law to surface 2. There is no charge inside the surface.

$$\Phi = \frac{\Sigma q_{inside}}{\epsilon_0}$$

$$\Phi_{right} + \Phi_{left} = 0$$

$$\vec{E}_{right} \cdot \hat{n}_{right} A_{box} + \vec{E}_{left} \cdot \hat{n}_{left} A_{box} = 0$$

$$< E_{right}, 0, 0 > \cdot < 1, 0, 0 > A_{box} + < E_{left}, 0, 0 > \cdot < -1, 0, 0 > A_{box} = 0$$

$$E_{right} A_{box} - E_{left} A_{box} = 0$$

$$E_{right} - E_{left} = 0$$

$$E_{right} = E_{left}$$

The electric therefore is uniform in this region since $E_{right} = E_{left}$. If this was not the case, then the flux through the box would not be zero and would violate Gauss' Law.

(c) For box 3, $\Phi_{total} = \Phi_{left}$ where Φ_{left} is the flux through the left side of the box. (The reasoning is similar to the reasoning in part (a)).

$$\Phi = \frac{\Sigma q_{inside}}{\epsilon_0}$$

$$-EA_{box} = \frac{Q_{box}}{\epsilon_0}$$

$$E = \frac{\frac{Q_{box}}{A_{box}}}{\epsilon_0}$$

Since the plate is uniformly charged, then $\frac{Q_{box}}{A_{box}} = \frac{Q}{A}$. Thus,

$$E = -\frac{\frac{Q}{A}}{\epsilon_0}$$

(d) For box 4, the only flux is the flux through the left side. $\vec{E} =< -E_{fringe}, 0, 0 >$ and $\hat{n} =< 1, 0, 0 >$. So, Gauss' Law gives

$$\Phi = \frac{\Sigma q_{inside}}{\epsilon_0}$$

$$E_{fringe} A_{box} = \frac{q_{box}}{\epsilon_0}$$

$$E_{fringe} = \frac{\frac{q_{box}}{A_{box}}}{\epsilon_0}$$

Since the charge is uniformly distributed, $\frac{q_{box}}{A_{box}} = q_A$, so $E_{fringe} = \frac{\frac{q}{A}}{\epsilon_0}$.

Substitute for E_{fringe} and solve for q.

$$E_{fringe} = \frac{\frac{q}{A}}{\epsilon_0}$$

$$\frac{\frac{Q}{A}}{2\epsilon_0} \frac{s}{R} = \frac{\frac{q}{A}}{\epsilon_0}$$

$$q = \frac{1}{2}\left(\frac{s}{R}\right) Q$$

As expected, $q << Q$ since $s << R$.

P25:
Solution:

(a) From the diagram in the figure below it is clear that there is cancellation of the vertical components of magnetic field contributed by two wires to the left and the right of the observation location. Therefore the direction of the magnetic field must be to the left at the location above the wires and to the right at the location below the wires.

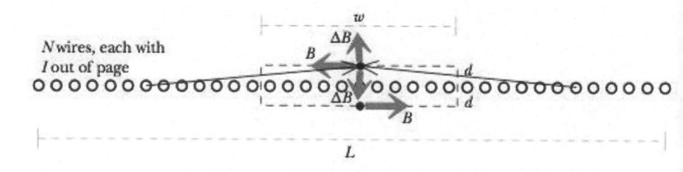

(b) Use Ampere's law, and go counterclockwise around the closed rectangular path.

Along the sides of the path $\int \vec{B} \cdot d\vec{l} = 0$, since $\vec{B}$ is perpendicular to $d\vec{l}$.

Along the upper part of the path, $\int \vec{B} \cdot d\vec{l} = Bw$.

Along the lower part of the path, $\int \vec{B} \cdot d\vec{l} = Bw$.

Therefore, applying Ampere's law gives

$$\int \vec{B} \cdot d\vec{l} = \mu_0 I_{\text{inside path}}$$
$$2Bw = \mu_0 I_{\text{inside path}}$$

The current inside the amperian loop is $I_{\text{inside path}} = \left(\frac{N}{L}\right) wI$ since there are N/L current-carrying wires per meter, and a width w of the enclosing path. Thus,

$$2Bw = \mu_0 \left(\frac{N}{L}\right) wI$$
$$B = \frac{\mu_0 NI}{2L}$$

(c) The magnetic field due to each sheet wires is in the same direction, so inside the two sheets of wires $B = \frac{\mu_0 NI}{L}$.

(d) In this case, the magnetic fields due to the sheets of wires are in opposite directions and subtract, so $B = 0$.

P27:
Solution:

Apply Ampere's law to the Amperian path. Looking at the path from the right end, integrate around the path counterclockwise. Then I_1 is flowing out of the surface and is positive. I_2 is flowing into the surface and is negative. $\vec{B}$ is tangent to the path in the direction of $d\vec{l}$, so $\oint \vec{B} \cdot d\vec{l} = B \oint d\vec{l} = B(2\pi R)$. Ampere's law gives

$$\oint \vec{B} \cdot d\vec{l} = \mu_0 I_{\text{inside path}}$$
$$B(2\pi R) = \mu_0 |I_1 - I_2|$$
$$B = \frac{\mu_0}{2\pi} \frac{|I_1 - I_2|}{R}$$

We can write this in a more familiar form by multiplying by 2/2 to give

$$B = \frac{\mu_0}{4\pi} \frac{2|I_1 - I_2|}{R}$$

This is the same as the magnitude of the magnetic field at a distance R from a long, straight wire with current $I_1 - I_2$. Note that if $I_1 = I_2$, then $B = 0$ as expected. Also, if $I_2 > I_1$, then B will be in the opposite direction meaning that $\vec{B}$ would be tangent to the path and clockwise.

CP30:
Solution:

According to Gauss' Law, if the charged particle is inside the cube (anywhere inside), then the total electric flux through the surface of the cube is

$$
\begin{aligned}
\Phi_{elec} &= \frac{\sum q_{inside}}{\varepsilon_0} \\
&= \frac{2 \times 10^{-9} \text{ C}}{8.85 \times 10^{-12} \text{ C}^2/(\text{N} \cdot \text{m}^2)} \\
&= 226 \text{ V} \cdot \text{m}
\end{aligned}
$$

In the program below, we have avoided using lists and other features of Python so that the program is transparent to a student who is learning Python in this physics course. Loops are needed in order to break each face of the cube into smaller squares and sum the electric flux through each square. A loop is needed for each axis of each face of the cube.

To increase the number of squares, change the variable `Nside` which represents the number of squares along an edge of the face of the cube. Thus, the total number of squares on each face is `Nside`^{2}.

Here are the results of the program for various numbers of squares. I rounded the electric flux to 3 significant figures since that is the number I used for ε_0.

squares per face	Φ_{elec} V $\cdot$ m
1	432
4	235
9	233
16	230
25	228
36	228
49	227
100	227
225	226

For more than 225 squares on each face of the cube, the electric flux does not change (for 3 significant figures).

When the charged particle is moved to a location outside the cube, the results are:

squares per face	Φ_{elec} V $\cdot$ m
9	-3×10^{-4}
100	-2.7×10^{-5}
10,000	-2.7×10^{-7}

Even for small numbers of squares, the result is very close to zero. The electric flux get smaller (closer to zero) for larger numbers of squares on each face.

Here is the sample program. Note that it prints the electric flux but does not display the charge nor the cube.

```
from __future__ import division, print_function
from visual import *

q=2e-9 #C
rq=vector(-0.03,0,0) #location of q
oofpez=9e9

L=0.01 #length of side of cube
Nside=1 #number of squares along a side of the cube
N=Nside*Nside #total number of squares for each face of the cube
```

```
rcube=vector(-0.03,0,0) #center of cube

#length of the side of a square
Lsquare=L/Nside
dx=Lsquare

#center of first square
x0=rcube.x-L/2+dx/2
y0=rcube.y-L/2+dx/2
z0=rcube.z-L/2+dx/2

#center of edge of last square
xf=rcube.x+L/2
yf=rcube.y+L/2
zf=rcube.z+L/2

#starting values of the coordinates of the center of a square
x=x0
y=y0
z=z0

#iteration number; this is needed to make sure we iterate at least once
i=1
j=1

flux=0
dA=dx*dx

#xy plane at -z
while x<=xf or i==1:
    while y<=yf or j==1:
        r=vector(x,y,rcube.z-L/2)
        rrel=r-rq
#        print(rrel)
        E=oofpez*q/mag(rrel)**2*norm(rrel)
#        print(E)
        dAvec=dA*vector(0,0,-1)
        flux=flux+dot(E,dAvec)
#        print(flux)
        y=y+dx
        j=j+1
    y=y0
    x=x+dx
    i=i+1
#print total flux
#print("The total flux is ", flux, " N/C m^2")

#reset starting coordinates and iteration number
x=x0
y=y0
z=z0
```

```
i=1
j=1

#xy plane at +z
while x<=xf or i==1:
    while y<=yf or j==1:
        r=vector(x,y,rcube.z+L/2)
        rrel=r-rq
        E=oofpez*q/mag(rrel)**2*norm(rrel)
        dAvec=dA*vector(0,0,1)
        flux=flux+dot(E,dAvec)
        y=y+dx
        j=j+1

    y=y0
    x=x+dx
    i=i+1

#reset starting coordinates and iteration number
x=x0
y=y0
z=z0

i=1
j=1
#xz plane at -y
while x<=xf or i==1:
    while z<=zf or j==1:
        r=vector(x,rcube.y-L/2,z)
        rrel=r-rq
        E=oofpez*q/mag(rrel)**2*norm(rrel)
        dAvec=dA*vector(0,-1,0)
        flux=flux+dot(E,dAvec)
        z=z+dx
        j=j+1

    z=z0
    x=x+dx
    i=i+1

#reset starting coordinates and iteration number
x=x0
y=y0
z=z0

i=1
j=1
#xz plane at +y
while x<=xf or i==1:
    while z<=zf or j==1:
        r=vector(x,rcube.y+L/2,z)
        rrel=r-rq
```

```
        E=oofpez*q/mag(rrel)**2*norm(rrel)
        dAvec=dA*vector(0,1,0)
        flux=flux+dot(E,dAvec)
        z=z+dx
        j=j+1

    z=z0
    x=x+dx
    i=i+1

#reset starting coordinates and iteration number
x=x0
y=y0
z=z0

i=1
j=1
#yz plane at +x
while y<=yf or i==1:
    while z<=zf or j==1:
        r=vector(rcube.x+L/2,y,z)
        rrel=r-rq
        E=oofpez*q/mag(rrel)**2*norm(rrel)
        dAvec=dA*vector(1,0,0)
        flux=flux+dot(E,dAvec)
        z=z+dx
        j=j+1

    z=z0
    y=y+dx
    i=i+1

#reset starting coordinates and iteration number
x=x0
y=y0
z=z0

i=1
j=1
#yz plane at -x
while y<=yf or i==1:
    while z<=zf or j==1:
        r=vector(rcube.x-L/2,y,z)
        rrel=r-rq
        E=oofpez*q/mag(rrel)**2*norm(rrel)
        dAvec=dA*vector(-1,0,0)
        flux=flux+dot(E,dAvec)
        z=z+dx
        j=j+1

    z=z0
    y=y+dx
```

```
    i=i+1

#print total flux
print("The total flux is ", flux, " N/C m^2")
print("q/epsilon_0 = ", 4*pi*oofpez*q, " N/C m^2")
```

22   Chapter 22: Faraday's Law

Q01:
Solution:

(a) $-y$ direction

(b) $-y$ direction

(c) B at the origin is larger at time t_2 because the magnet is closer to the origin.

(d) $-y$ direction

(e) $+y$ direction

(f) Counterclockwise. Thumb points upward, fingers curl counterclockwise.

(g) $\vec{E}_{NC}$ is in the $-z$ direction at points on the $+x$ axis.

Q07:
Solution:

$\Delta\vec{B}$ is upward ($+y$ direction), so $-\Delta\vec{B}$ is downward. Using the right-hand rule, the induced current flows clockwise around the loop if viewed from above.

P13:
Solution:

The magnitude of the average emf around the coil is

$$
\begin{aligned}
|\text{emf}| &= \left|\frac{\Delta\Phi}{\Delta t}\right| \\
&= \frac{0.3\ \text{T}\cdot\text{m}^2}{0.2\ \text{s}} \\
&= 1.5\ \text{V}
\end{aligned}
$$

$\Delta\vec{B}$ is to the left, toward the magnet. $-\Delta\vec{B}$ is toward the right, away from the magnet. With your thumb pointed to the right, your fingers curl counterclockwise around the loop, as viewed from the right side of the coil, so current will flow counterclockwise as viewed from the right side of the coil.

P17:
Solution:

(a)

$$
\begin{aligned}
\left|\Phi_{\text{mag}}\right| &= \left|\vec{B}\right|A = \left(\frac{\mu_o}{4\pi}\right)A \approx \left(\frac{\mu_o}{4\pi}\frac{2NI\pi r_c^3}{d_{cl}^3}\right)\left(\pi r_l^2\right) \\
&\approx \left(1\times 10^{-7}\ \frac{\text{T}\cdot\text{m}^2}{\text{C}\cdot\text{m/s}}\right)\frac{2(300)(5\ \text{A}\pi^2\left(0.09\ \text{m}\right)^2\left(0.04\ \text{m}\right)^2}{\left(0.22\ \text{m}\right)^3} \\
&\approx 2.6\times 10^{-6}\ \text{T}\cdot\text{m}^2
\end{aligned}
$$

(b) Treat the coil as a magnetic dipole.

(c) There is no electric field in the loop.

(d) $\frac{dI}{dt} = -0.3$ A/s and $\vec{E}_{NC}$ is in the $-y$ direction.

(e)

$$
\begin{aligned}
\left| \frac{d\Phi_{mag}}{dt} \right| &= \frac{\mu_o}{4\pi} \frac{2N\pi^2 r_c^2 r_l^2}{d_{cl}^3} \left| \frac{dI}{dt} \right| \\
&\approx \left(1 \times 10^{-7} \, \frac{\text{T} \cdot \text{m}^2}{\text{C} \cdot \text{m/s}} \right) \frac{2(300)\pi^2 (0.09 \text{ m})^2 (0.04 \text{ m})^2}{(0.22 \text{ m})^3} (0.3 \text{ A/s}) \\
&\approx 2.16 \times 10^{-7} \text{ V}
\end{aligned}
$$

(f)

$$
\begin{aligned}
|\text{emf}| &= \left| \frac{d\Phi_{mag}}{dt} \right| \\
&\approx 2.16 \times 10^{-7} \text{ V}
\end{aligned}
$$

(g)

$$
\begin{aligned}
|\text{emf}| &= \oint_C \vec{E}_{NC} \bullet d\vec{l} = 2\pi r_l \left| \vec{E}_{NC} \right| \\
\left| \vec{E}_{NC} \right| &= \frac{|\text{emf}|}{2\pi r_l} \\
&\approx \frac{2.16 \times 10^{-7} \text{ V}}{2\pi (0.04 \text{ m})} \\
&\approx 8.59 \times 10^{-7} \text{ V/m}
\end{aligned}
$$

(h) Removing the loop doesn't change the curly electric field.

P21:
Solution:

(a) $\vec{B}$ is out of the page (toward the magnet) and is increasing. Thus, $\frac{d\vec{B}}{dt}$ is out of the page and $-\frac{d\vec{B}}{dt}$ is into the page. So $\vec{E}$ at locations in the coil curls clockwise, if you are facing the coil. At point 1, $\vec{E}_1$ points upward and at point 2, $\vec{E}_2$ points downward.

(b) We will approximate the emf by calculating the average emf using Faraday's Law.

$$
\begin{aligned}
|\text{emf}| &\approx \left| \frac{\Delta I}{\Delta t} \right| \\
&\approx N_{coil} A_{coil} \left| \frac{\Delta B}{\Delta t} \right|
\end{aligned}
$$

Note that B due to the magnet is given by $B = \frac{\mu_o}{4\pi} \frac{2\mu}{x^3}$ so you can calculate $|\Delta B|$ during the process.

$$B = \frac{\mu_o}{4\pi}\frac{2\mu}{x^3}$$

$$B_i = (1 \times 10^{-7}\,\frac{\text{T}\cdot\text{m}}{\text{A}})\frac{2(0.8\,\text{A}\cdot\text{m}^2)}{(0.4\,\text{m})^3} = 2.5 \times 10^{-6}\,\text{T}$$

$$B_f = (1 \times 10^{-7}\,\frac{\text{T}\cdot\text{m}}{\text{A}})\frac{2(0.8\,\text{A}\cdot\text{m}^2)}{(0.3\,\text{m})^3} = 5.93 \times 10^{-6}\,\text{T}$$

$$\Delta B = B_f - B_i = 3.43 \times 10^{-6}\,\text{T}$$

So the everage emf is

$$
\begin{aligned}
|\text{emf}| &\approx N_{\text{coil}}A_{\text{coil}}\frac{|\Delta B|}{\Delta t} \\
&\approx 3000\pi(0.05\,\text{m}^2)\frac{3.43 \times 10^{-6}\,\text{T}}{0.2\,\text{s}} \\
&\approx 4.04 \times 10^{-4}\,\text{V} \\
&\approx 0.404\,\text{mV}
\end{aligned}
$$

(c) We assumed that B was uniform across the plane of the coil. Also, we calculated the average emf. In fact, at $x = 40$ cm, the emf is less than this value, and at $x = 30$ cm, the emf is greater than this value (assuming the bar magnet is moved at a constant speed).

Note that the emf can be calculated at a given location x, if you wish.

$$
\begin{aligned}
B &= \frac{\mu_o}{4\pi}\frac{2\mu}{x^3} \\
\left|\frac{dB}{dt}\right| &= \left|\frac{\mu_o}{4\pi}\frac{2\mu(-3)}{x^4}\frac{dx}{dt}\right| \\
&= \frac{\mu_o}{4\pi}\frac{6\mu}{x^4}v
\end{aligned}
$$

$$
\begin{aligned}
|\text{emf}| &= \left|\frac{d\Phi}{dt}\right| \\
&= N_{\text{coil}}A_{\text{coil}}\left|\frac{dB}{dt}\right| \\
&= N_{\text{coil}}A_{\text{coil}}\frac{\mu_o}{4\pi}\frac{6\mu}{x^4}v
\end{aligned}
$$

Use this to calculate emf at any x. Note: $v = \frac{0.1\,\text{m}}{0.2\,\text{s}} = 0.5$ m/s.

At $x = 0.4$ m, emf $= 2.21 \times 10^{-4}$ V.

At $x = 0.3$ m, emf $= 6.98 \times 10^{-4}$ V.

Note that the arithmetic mean is not the same as the average (over time) since emf varies as $\frac{1}{x^4}$. However, you can see that the emf at $x = 0.4$ m is less than emf_{ave}, and the emf at $x = 0.3$ m is greater than emf_{ave}.

P25:
 Solution:

(a) $\vec{B}_1$ at the location of coil 2 is decreasing and is in the -z direction. So, $\frac{d\vec{B}}{dt}$ is in the +z direction, and $-\frac{d\vec{B}}{dt}$ is in the -z direction.

(b) As viewed from the +z axis, $\vec{E}$ curls closkwise around coil 2. Thus at the top of the coil, $\vec{E}$ is in the +x direction.

(c)

$$
\begin{aligned}
\Phi &= BA\cos 0^\circ \\
&= \left(\frac{\mu_o}{4\pi}\frac{2I_1 A_1 N_1}{(z^2+R_1^2)^{\frac{3}{2}}}\right)A_2 \\
&= \frac{\mu_o}{4\pi}\frac{2(18\text{ A})\pi(0.07\text{ m})^2(570)\pi(0.03\text{ m})^2}{((0.14\text{ m})^2+(0.07\text{ m})^2)^{\frac{3}{2}}} \\
&= 2.33\times10^{-5}\text{ T}\cdot\text{m}^2
\end{aligned}
$$

(d) We approximated the coils as thin coils. We assumed $\vec{B}$ was uniform across the plane of coil 2.

(e) At $t=0.4$ s, $I_1=6$ A and $\Phi=\frac{1}{3}\Phi_{t=0}=7.76\times10^{-6}$ T$\cdot$m^2. Note that $\Phi\propto I_1$, so since $I_1=\frac{1}{3}I_{t=0}$, then the magnetic flux decreases by the same factor.

(f)

$$
\begin{aligned}
|\text{emf}| &= \left|\frac{d\Phi}{dt}\right| = \frac{|7.76\times10^{-6}\text{ T}\cdot\text{m}^2 - 2.33\times10^{-5}\text{ T}\cdot\text{m}^2|}{0.4\text{ s}} \\
&= 3.88\times10^{-5}\text{ V}
\end{aligned}
$$

(g)

$$
\begin{aligned}
\Delta V &= N|\text{emf}|_{1\text{ turn}} \\
&= 275(3.88\times10^{-5}\text{ V}) \\
&= 0.0107\text{ V} \\
&= 10.7\text{ mV}
\end{aligned}
$$

(h) Since $\vec{E}$ is uniform inside the wire, $EL=\Delta V$, where $L=N(2\pi R_2)$. So,

$$
\begin{aligned}
E &= \frac{\Delta V}{L} \\
&= \frac{\Delta V}{N2\pi R_2} \\
&= \frac{0.0107\text{ V}}{(275)2\pi(0.03\text{ m})} \\
&= 2.06\times10^{-4}\frac{\text{V}}{\text{m}}
\end{aligned}
$$

(i) 1 and 4 are true.

If I is constant, then B is constant and Φ is constant and $\frac{d\Phi}{dt}=0$. So, emf $=0$.

P33:

   Solution:

To visualize the problem, sketch the magnet from a top-view, looking downward in the $-y$ direction. A sketch of the situation is shown below. I have drawn a loop a distance 1 m (in the $-z$ direction) from the dipole. The radius of the loop is 0.04 m.

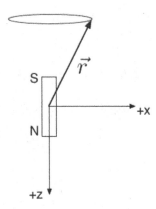

The vector $\vec{r}$ from the center of the dipole to the point where the electric field is calculated is:

$$
\begin{aligned}
\vec{r} &= \vec{r}_E - \vec{r}_{dipole} \\
&= <1.04, 5, 2> \text{ m} - <1, 5, 3> \text{ m} \\
&= <0.04, 0, -1> \text{ m}
\end{aligned}
$$

Apply Faraday's law to the loop.

$$
\begin{aligned}
|\text{emf}_{loop}| &= \left| \frac{d\Phi_{mag}}{dt} \right| \\
\oint \vec{E} \cdot d\vec{l} &= \left| \frac{d}{dt} \left(\int \vec{B} \cdot \hat{n} dA \right) \right|
\end{aligned}
$$

The magnetic field of a dipole at a distance z along the axis of the dipole has a magnitude

$$
B = \frac{\mu_o}{4\pi} \frac{2\mu}{z^3}
$$

The magnetic field due to the dipole is parallel to $\hat{n}$ for the loop. Using the approximation that B is uniform across the plane of the loop, then the magnetic flux through the loop is

$$
\begin{aligned}
\Phi_{mag} &= \int \vec{B} \cdot \hat{n} dA \\
&= \pi r^2 B
\end{aligned}
$$

where r is the radius of the loop and B is the magnetic field at the location of the loop.

The emf around the loop is

$$\oint \vec{E} \cdot d\vec{l} \;=\; E2\pi r$$

Now substitute the emf and magnetic flux into Faraday's law.

$$
\begin{aligned}
\oint \vec{E} \cdot d\vec{l} &= \left| \frac{d}{dt}\left(\int \vec{B} \cdot \hat{n}\,dA \right) \right| \\
E2\pi r &= \left| \frac{d}{dt}(\pi r^2 B) \right| \\
&= \left| \pi r^2 \frac{dB}{dt} \right| \\
&= \left| \pi r^2 \frac{\mu_o}{4\pi} 2\mu \frac{-3}{z^4}\frac{dz}{dt} \right| \\
&= \left| \pi r^2 \frac{\mu_o}{4\pi} 2\mu \frac{-3}{z^4} v_z \right|
\end{aligned}
$$

Now solve for the magnitude of the electric field.

$$
\begin{aligned}
E &= \frac{\mu_o}{4\pi}\frac{3\mu r v}{z^4} \\
&= (1\times10^{-7}\ \text{T}\cdot\text{m/A})\frac{3(6\ \text{A}\cdot\text{m}^2)(0.04\ \text{m})(5\ \text{m/s})}{(1\ \text{m})^4} \\
&= 3.6\times10^{-7}\ \text{N/C}
\end{aligned}
$$

The direction of the electric field is given by Lenz's Law. The magnetic field at the location $\vec{r}$ is in the $+z$ direction and is decreasing (because the magnet is moving away from this point). As a result, $-\Delta\vec{B}$ is in the $+z$ direction. Using the right-hand rule with your thumb pointing in the $+z$ direction, $\vec{E}$ curls counterclockwise around the $+z$ axis. Thus the electric field at $\vec{r}$ is in the $+y$ direction, and $\vec{E} = <0, 3.6\times10^{-7}, 0>$ N/C.

P39:
Solution:

For a transformer,

$$\frac{\text{emf}_1}{N_1} = \frac{\text{emf}_2}{N_2}$$

Solve for emf_2, the emf across the secondary coil.

$$\begin{aligned}
\text{emf}_2 &= (N_2/N_1)\text{emf}_1 \\
&= \frac{350}{100}(120 \text{ V}) \\
&= (3.5)(120 \text{ V}) \\
&= 420 \text{ V}
\end{aligned}$$

The above relationship for emf_1 and emf_2 is true for all times. Though emf_1 varies in time, at any given instant, they are related by the above equation.

Because energy is conserved, $P_1 = P_2$. Since $P = I\Delta V$ then

$$\begin{aligned}
I_2\Delta V_2 &= I_1\Delta V_1 \\
I_2 &= I_1\frac{\Delta V_1}{\Delta V_2} \\
&= (4 \text{ A})\left(\frac{120}{420}\right) \\
&= 1.1 \text{ A}
\end{aligned}$$

23   Chapter 23: Electromagnetic Radiation

Q01:
   Solution:

   (a) (d) Ampere-Maxwell law

   (b) (c) Faraday's law

   (c) (a) Gauss's law

   (d) (b) Gaus's law for magnetism

Q05:
   Solution:

   (a) a

   (b) c

   (c) g

   (d) j

Q09:
   Solution:

   Simply put, the direction of the force due to radiation pressure on a positively charged particle is in the same direction as for a negatively charged particle. As a result, the net radiation pressure on the neutral dust grain will be to the right.

   Model a neutral atom in the particle of dust as a positively charged nucleus with a negatively charged electron cloud surrounding the nucleus, so it can be treated as an electric dipole. Suppose that a pulse of electromagnetic radiation travels to the right in the $+x$ direction, with $\vec{E}$ upward in the $+y$ direction and $\vec{B}$ outward in the $+z$ direction. When this pulse reaches the neutral atom, the electric force on the positively charged nucleus is upward, causing the nucleus to accelerate upward. With an upward velocity, the magnetic field exerts a force on the nucleus that is to the right, in the $+x$ direction. Now consider the electron cloud. The electric field exerts a downward force on the electron cloud accelerating it downward and giving it a downward velocity. The magnetic force by the magnetic field on the downward moving electron cloud is also to the right. Thus, the radiation pressure on the neutral atom is to the right.

Q15:
   Solution:

   (2) The emerging beam bends away from the normal. Let 1 be the higher index medium and 2 be the lower index medium.

$$
\begin{aligned}
n_1 \sin\theta_1 &= n_2 \sin\theta_2 \\
\sin\theta_2 &= \frac{n_1}{n_2}\sin\theta_1
\end{aligned}
$$

   If $n_1 > n_2$, then $\sin\theta_2 > \sin\theta_1$, and $\theta_2 > \theta_1$.

P17:

Solution:

Traverse the loop in the counterclockwise direction so positive change in electric flux is out of the page. Use Maxwell's version of Ampére's law, and note there are no charged particles so the conventional current term is zero.

$$
\oint_C \vec{B} \bullet d\vec{l} = \mu_o \varepsilon_o \frac{d\Phi_E}{dt}
$$

$$
= \int_{\text{left}} \vec{B} \bullet d\vec{l} + \int_{\text{right}} \vec{B} \bullet d\vec{l}
$$

$$
-\left|\vec{B}_l\right| L - \left|\vec{B}_r\right| L = \frac{1}{c^2} L W \left|\frac{d\vec{E}}{dt}\right|
$$

$$
\left|\frac{d\vec{E}}{dt}\right| = \frac{c^2}{W}\left(\left|\vec{B}_l\right| + \left|\vec{B}_r\right|\right)
$$

$$
\approx \frac{\left(3 \times 10^8 \, \frac{m}{s}\right)^2}{4 \times 10^{-2} \, m}(0.32 \, \text{T} + 0.25 \, \text{T})
$$

$$
\approx 1.3 \times 10^{18} \, \frac{\text{N/C}}{\text{s}}
$$

We can conclude that there is an electric field **changing** into the page inside the rectangular loop. In other words, there is effectively a current into the page through the loop.

P21:

Solution:

The direction of propagation is in the direction of $\vec{E} \times \vec{B}$ which is in the $-z$ direction. Using the right-hand rule, $\vec{B}$ must be in the $-y$ direction. Its magnitude is given by

$$
E = cB
$$

$$
B = \frac{E}{c}
$$

$$
= \frac{7.2 \times 10^6 \, \text{N/C}}{3 \times 10^8 \, \text{m/s}}
$$

$$
= 0.024 \, \text{T}
$$

P25:

Solution:

(a) The wave propagates at the speed of light. So:

$$
c = \frac{distance}{\Delta t}
$$

$$
\Delta t = \frac{distance}{c}
$$

$$
= \frac{0.15 \, \text{m}}{3 \times 10^8 \, \frac{m}{s}}
$$

$$
= 5.00 \times 10^{-10} \, \text{s}
$$

(b) For a proton, the radiative electric field is opposite $\vec{a}_\perp$ which is in the $-y$ direction. Thus, at the given location, $\left|\vec{E}\right|_{radiative}$ is in the $+y$ direction.

(c) For an electron, the radiative electric field is in the same direction as $\vec{a}_\perp$. Thus, at the given location, $\left|\vec{E}\right|_{radiative}$ is in the $-y$ direction.

P29:

  Solution:

(a) At $t_2 = 1$ ns the only electric field is the Coulomb field to the left, of magnitude

$$E = \frac{1}{4\pi\varepsilon_o}\frac{e}{r^2} = \left(9 \times 10^9\ \frac{\text{N}\cdot\text{m}^2}{\text{C}^2}\right)\frac{1.602 \times 10^{-19}\ \text{C}}{(15\ \text{m})^2} = 6.4 \times 10^{-12}\ \text{N/C}$$

(b) The radiative electric is first observed at time t_3 where:

$$t_3 = \frac{15\ \text{m}}{3 \times 10^8\ \frac{\text{m}}{\text{s}}} = 5 \times 10^{-8}\ \text{s} = 50\ \text{ns}$$

(c) The radiative electric field is proportional to $-q\vec{a}_\perp = +e\vec{a}_\perp$, which is upward. So starting at $t_3 = 50$ ns, we observe a radiative field whose direction is upward. There is an accompanying magnetic field out of the page because the direction of propagation $\vec{E} \times \vec{B}$ is to the right.

(d) The magnitude of the upward radiative field is

$$\begin{aligned}
\left|\vec{E}\right|_{radiative} &= \frac{1}{4\pi\varepsilon_o}\frac{ea_\perp}{c^2 r} \\
&= \left(9 \times 10^9\ \frac{\text{N}\cdot\text{m}^2}{\text{C}^2}\right)\frac{(1.602 \times 10^{-19}\ \text{C})(1 \times 10^{18}\ \text{m/s}^2)}{(3 \times 10^8\ \frac{\text{m}}{\text{s}})^2(15\ \text{m})} \\
&= 1.07 \times 10^{-9}\ \text{N/C}
\end{aligned}$$

This is so much larger than the Coulomb field to the left that the electric field at location A is almost solely just the upward radiative field.

(e) A positive charge is accelerated upward by the upward radiative electric field. Now that there is an upward velocity, the magnetic field out of the page exerts a force to the right (radiation pressure).

P35:

  Solution:

  radio, $f = 100$ kHz:

$$c = \lambda f$$
$$\lambda = \frac{c}{f}$$
$$= \frac{3 \times 10^{8} \frac{m}{s}}{1 \times 10^{6} \text{ Hz}}$$
$$= 300 \text{ m}$$

television, $f = 100$ MHz:

$$c = \lambda f$$
$$\lambda = \frac{c}{f}$$
$$= \frac{3 \times 10^{8} \frac{m}{s}}{1 \times 10^{8} \text{ Hz}}$$
$$= 3 \text{ m}$$

red light, $f = 4.3 \times 10^{14}$ Hz:

$$c = \lambda f$$
$$\lambda = \frac{c}{f}$$
$$= \frac{3 \times 10^{8} \frac{m}{s}}{4.3 \times 10^{14} \text{ Hz}}$$
$$= 6.98 \times 10^{-7} \text{ m}$$
$$= 698 \text{ nm}$$

blue light, $f = 7.5 \times 10^{14}$ Hz:

$$c = \lambda f$$
$$\lambda = \frac{c}{f}$$
$$= \frac{3 \times 10^{8} \frac{m}{s}}{7.5 \times 10^{14} \text{ Hz}}$$
$$= 4.00 \times 10^{-7} \text{ m}$$
$$= 400 \text{ nm}$$

P43:
Solution:

(a) You need to know the angle of the light beam as it interacts with the water. Use the triangle and geometry and remember to define θ_1 as the angle of the light beam with respect to the vertical.

$$\tan \theta_1 = \frac{2.4 \text{ m}}{1.2 \text{ m}}$$
$$= 2$$
$$\theta_1 = 63.4°$$

(b) Use Snell's law to calculate the angle that the beam emerges from the air in the water.

$$n_1 \sin \theta_1 = n_2 \sin \theta_2$$
$$\sin \theta_2 = \frac{n_1}{n_2} \sin \theta_1$$
$$= \frac{1.00029}{1.33} \sin 63.9°$$
$$\theta_2 = 42.3°$$

Note that it bent toward the normal, as expected.

(c) Now use geometry to find the horizontal distance traveled *from the point where the beam entered the water* to the point where it hits the bottom of the pool.

$$\tan \theta_2 = \frac{\Delta x}{2.0 \text{ m}}$$
$$\Delta x = (2.0 \text{ m}) \tan(42.3°)$$
$$= 1.82 \text{ m}$$

The total distance from the edge of the pool is $2.4 \text{ m} + 1.82 \text{ m} = 4.22 \text{ m} \approx 4.2 \text{ m}$.

P47:
Solution:

Sketch a picture of the lens and the object and sketch a few rays from the object. The easiest rays to sketch are (1) the ray that passes through the center of the lens unrefracted, (2) the ray that travels parallel to the optic axis and refracts through the focus, and (3) the ray that travels through the focus and refracts parallel to the optic axis.

Apply the thin-lens formula to get the x-coordinate of the point source:

$$\frac{1}{f} = \frac{1}{d_1} + \frac{1}{d_2}$$
$$\frac{1}{d_2} = \frac{1}{f} - \frac{1}{d_1}$$
$$d_2 = \left(\frac{1}{0.2 \text{ m}} - \frac{1}{0.25 \text{ m}} \right)^{-1}$$
$$= 1.0 \text{ m}$$

The image is located 1.0 m from the lens, on the opposite side of the lens from the object. The image is a real image because it is formed by converging rays on the opposite of the lens from the object.

The vertical position is magnified and inverted, so the y-coordinate will be below the optic axis and will have a magnitude greater than 0.01 m. If you analyze similar triangles using the ray that passes through the center of the lens and the optic axis, then

$$\frac{y_2}{d_2} = -\frac{y_1}{d_1}$$

$$y_2 = -\frac{y_1}{d_1}d_2$$

$$= -\frac{0.01 \text{ m}}{0.25 \text{ m}}(1.0 \text{ m})$$

$$= -0.04 \text{ m}$$

So the image of the point source will be 4 cm below the optic axis. The figure below shows the image for a point source of light as the object on the left side of the lens. It is at $< 100, -4, 0 >$ cm.

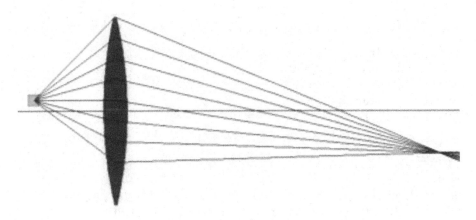

CP51:
Solution:

(a) Here is a sample VPython program. A scaling factor is used for the electric field arrows as well as the magnetic field arrows so that they can be clearly displayed.

```
from __future__ import division
from visual import *

scene.width=1000
scene.height=1000
scene.x = scene.y = 0
scene.background = color.white

c = 3e8
lamb = 600
omega = 2*pi*c/lamb
scene.range=3*lamb

xx = arange(-3*lamb,3.001*lamb, lamb/20)

xhat = vector(1,0,0)

Evec = []
z=0
y=0
for x in xx:
```

```
        ea = arrow(pos=(x,y,z), axis=(0,lamb/10,0), color=(1,.6,0),
                   shaftwidth=lamb/45, fixedwidth=1)
        ba = arrow(pos=(x,y,z), axis=(0,0,0), color=(0,1,1),
                   shaftwidth=lamb/45, fixedwidth=1)
        ea.B = ba
        Evec.append(ea)

t = 0
dt = lamb/c/100
E0 = 1e4
B0 = E0/c
Escale = lamb/2/E0
Bscale = c*Escale

print 'dt (s) = ',dt
print 'wavelength (m) = ', lamb
print 'omega (rad/s) = ', omega
print 'f (Hz) = ', omega/2/pi
print 'E_0 (V/m) ', E0

while 1:
    rate(100)
    t = t+dt
    for ea in Evec:
        E=vector(0,E0*cos(omega*t - 2*pi/lamb*ea.x),0)
        ea.axis = Escale*E
        Bmag=mag(E)/c
        Bdir=cross(xhat,norm(E))
        B=Bmag*Bdir
        ea.B.axis = B*Bscale
```

(b) An example program is shown below. Note that the time step dt should be very small in order to reduce numerical error.

```
from __future__ import division
from visual import *

scene.width=1000
scene.height=1000
scene.x = scene.y = 0
scene.background = color.white

c = 3e8
lamb = 600
omega = 2*pi*c/lamb
scene.range=3*lamb

positron=sphere(pos=(-3*lamb,0,0), radius=lamb/10, color=color.red)
positron.m=9.11e-31
positron.q=1.6e-19
positron.v=vector(0,0,0)
positron.p=positron.m*positron.v/sqrt(1-mag(positron.v)**2/c**2)
positron.trail = curve(color=positron.color)
```

```
xx = arange(-3*lamb,3.001*lamb, lamb/20)

xhat = vector(1,0,0)

Evec = []
z=0
y=0
for x in xx:
    ea = arrow(pos=(x,y,z), axis=(0,lamb/10,0), color=(1,.6,0),
               shaftwidth=lamb/45, fixedwidth=1)
    ba = arrow(pos=(x,y,z), axis=(0,0,0), color=(0,1,1),
               shaftwidth=lamb/45, fixedwidth=1)
    ea.B = ba
    Evec.append(ea)

t = 0
dt = lamb/c/1000
E0 = 1e4
B0 = E0/c
Escale = lamb/2/E0
Bscale = c*Escale

print 'dt (s) = ',dt
print 'wavelength (m) = ', lamb
print 'omega (rad/s) = ', omega
print 'f (Hz) = ', omega/2/pi
print 'E_0 (V/m) ', E0


while 1:
    rate(100)
    t = t+dt

    #update the wave
    for ea in Evec:
        E=vector(0,E0*cos(omega*t - 2*pi/lamb*ea.x),0)
        ea.axis = Escale*E
        Bmag=mag(E)/c
        Bdir=cross(xhat,norm(E))
        B=Bmag*Bdir
        ea.B.axis = B*Bscale

    #calculate E, B, Fnet, and update the positron's momentum, velocity, and
        position
    E = vector(0,E0*cos(omega*t - 2*pi/lamb*positron.pos.x),0)
    Bmag = mag(E)/c
    Bdir = cross(xhat,norm(E))
    B = Bmag*Bdir
    Fnet = positron.q*E + positron.q*cross(positron.v,B)
    positron.p = positron.p + Fnet*dt
    positron.v = positron.p/positron.m*(1/sqrt(1+(mag(positron.p)/(positron.m*c))
```

```
        **2))
    positron.pos = positron.pos + positron.v*dt
    positron.trail.append(pos=positron.pos)
```